The Call of the Eco-Weird in Fiction, Films, and Games

"Who, or what, calls to us to take up the weird? The Call of the Eco-Weird names the weird as an urgent mode of ecocritical analysis in the paradoxical form of a call from an unknown, an elsewhere, an outside. Veering crucially away from methods of mastery or epistemological certainties about the world and instead towards the speculative and the always unresolved, the collection proposes weirdness as a primary way of engaging the complex and unprecedented confluence of ecological crises of the 21st century. Onishi and Bell have put together a valuable resource towards broader weird studies that demonstrates the eco-weird as intimately tied up with the most pressing currents of contemporary theory about the environment."

—Alison Sperling, *Assistant Professor of English, Florida State University*

"The Eco-Weird is a relatively new discipline in the interpretation of horror and supernatural fiction, but it is one that is much needed as we now face the dire consequences of global climate change. The authors in this book discuss some of the leading authors of weird fiction—H. P. Lovecraft, Algernon Blackwood, William Hope Hodgson, among others—and study how they addressed the conflict of humanity and nature. Films, role-playing games, and other media are also brought into the discussion. Overall, this book illuminates our tenuous position on a planet whose future is by no means assured."

—S. T. Joshi, author of Unutterable Horror: A History of Supernatural Fiction (2012)

Brian Hisao Onishi • Nathan M. Bell
Editors

The Call of the Eco-Weird in Fiction, Films, and Games

palgrave
macmillan

Editors
Brian Hisao Onishi
Department of Philosophy
Penn State Altoona
Altoona, PA, USA

Nathan M. Bell
Department of Philosophy and Religion
Dallas College
Dallas, TX, USA

ISBN 978-3-031-77125-5 ISBN 978-3-031-77126-2 (eBook)
https://doi.org/10.1007/978-3-031-77126-2

This Palgrave Macmillan imprint is published by the registered company Springer Nature Switzerland AG.
The registered company address is: Gewerbestrasse 11, 6330 Cham, Switzerland

Acknowledgments

The concept of the eco-weird has been developing over the past five years. We have had numerous levels of feedback and want to thank all those who have given their time and energy to any part of this project, including our eco-weird panel participants at the International Association of Environmental Philosophy. We want to thank Nicholas Mizer, Chloé Germaine, and Jon Newell for being a part of the early "Eco-Weird Worlds" study group. Your ideas have helped us to grow beyond our COVID-19 Zoom setting. We also want to thank everyone who participated in the inaugural meeting for the Society for the Study of the Eco-Weird, including Jonathan Beever and Alison Sperling. Your comments and engagement at that meeting offered a foundation for this text. We want to offer a special thanks to Heather Cole and the S.T. Joshi Endowed Research Fellowship housed at the John Hay Library of Brown University for offering partial funding and sponsorship for the research done in this volume. Finally, we want to thank all of our contributors. Your innovative ideas, helpful collaboration, and brilliant analysis have made this possible.

Contents

Notes on Contributors

Emily Alder is a Lecturer at Edinburgh Napier University, UK. She has written widely on the Nautical Gothic and Weird fiction. Her book, *Weird Fiction and Science at the Fin de Siècle*, was published by Palgrave Macmillan. Her publications have been featured in *Gothic Studies*, *Textual Practice*, and *Shima*. Her most recent article, "Arctic Ghosts: Whale Hunting and Haunting in Arthur Conan Doyle's 'The Captain of the Pole-Star'" will be published in *Victorian Studies*.

Nathan M. Bell is a Lecturer in Philosophy at Dallas College, USA. His research focuses on the intersection of philosophical hermeneutics and environmental philosophy. His work has been published at the forefront of the developing discipline of environmental hermeneutics, including a chapter forthcoming in *Environmental Hermeneutics in the Anthropocene: Nature and the Conflict of Interpretations.*

Chloé Germaine is a Reader in the Department of English at Manchester Metropolitan University, UK, and the co-director of the Manchester Game Centre. Chloé's research is focused at the intersection of game studies, gothic literature, and the Weird. Her work has been published in the journals *Analog Game Studies*, *Fantastika*, *Feminist Media Studies*, *Gothic Nature*, and *Horror Studies*. Her recent books, *The Dark Matter of Children's 'Fantastika' Literature: Speculative Entanglements* and *Material Game Studies: A Philosophy of Analogue Play*, were both published by Bloomsbury.

Dawn Keetley is a Professor of English and Film at Lehigh University, USA. She has written widely on folk-horror, horror film, and gothic fiction. She has edited numerous books, including *Jordan Peele's Get Out: Political Horror*, *Ecogothic in Nineteenth-Century American Literature* (with Matthew Sivils), *Plant Horror: Approaches to the Monstrous Vegetal in Fiction and Film* (with Angela Tenga), and *Folk Horror: New Global Pathways* (with Ruth Heholt). She has recently written a short book in the Cambridge Gothic Elements series called *Folk Gothic*. She also runs "Horror Homeroom," a website and podcast that discusses "horror in all its forms."

Nicholas J. Mizer is an Assistant Professor in the Games and Simulation Arts and Sciences Program at Rensselaer Polytechnic Institute, USA. He is the author of *Tabletop Role-Playing Games and the Experience of Imagined Worlds* and his writings on tabletop role-playing games have appeared in works such as *Role-Playing Game Studies: Transmedia Foundations*, *The International Encyclopedia of Anthropology*, and *Analog Game Studies*. He currently resides along the Machicantuck River in the traditional and unceded territory of the Haudenosaunee Confederacy.

Jonathan Newell is a faculty member in the English Department at Langara College, Vancouver, British Columbia. His research focuses on aesthetics and metaphysics in Weird fiction, horror, and the gothic. His book, *A Century of Weird Fiction, 1832–1937: Disgust, Metaphysics, and the Aesthetics of Cosmic Horror*, was published by the University of Wales Press in 2020.

Brian Hisao Onishi is an Associate Professor of Philosophy and Environmental Studies at Penn State Altoona, USA. His research focuses on the intersection between environmental philosophy and continental thought, with particular emphasis on wonder and Weird fiction. His work has appeared in *Environmental Philosophy*, *The Heythrop Journal*, and *SATS Northern European Journal of Philosophy*. His book, *Weird Wonder in Merleau-Ponty, Object-Oriented Ontology, and New Materialism*, was published by Palgrave Macmillan in 2023.

Brad Tabas is a philosopher. He is an Associate Professor in the Department of Social and Human Sciences at the École Nationale Supérieure de Techniques Avancées de Bretagne (ENSTA Bretagne), France. His recent work focuses on the philosophical, ecological, political, aesthetic, and ethical dimensions of the process of planetization. His recent work has been featured in *The Journal of Cavellian Studies*, *Cosmos and History: The Journal of Natural and Social Philosophy*, *Resilience*, and *International Journal of the Commons*.

Gry Ulstein works as a policy advisor at the Centre for Literary and Intermedial Crossings (CLIC), Vrije Universiteit Brussel (VUB), Brussels, Belgium. She has written widely on the intersection of Weird fiction and environmental issues. Her work has appeared in *More Posthuman Glossary*, *Sub-Stance*, and *Concentric*. Her most recent article, "The Weird and the Meta in Jeff VanderMeer's Dead Astronauts" was published in *Configurations* in 2022.

List of Figures

CHAPTER 1

Introduction

Brian Hisao Onishi and Nathan M. Bell

Ecohorror. EcoGothic. Ecosublime. Eco-criticism. Eco-phenomenology. Eco-weird. In a crowd of eco-driven subgenres, categories, and disciplines, the first question we must ask is whether we really need another cut in the academic fabric of environmental thought. Do we need another "eco" anything? This book offers a resounding yes to that question and provides an initial investigation of what we are calling the eco-weird. As we see it, the eco-weird maps the intersection of weird fiction, film, games, narratives, and environmental thought broadly construed. More than merely attending to a subgenre of narrative storytelling, we argue that the eco-weird is also a disciplinary approach to academic research.

Like the abovementioned genres and flavors of academic analysis, the "eco" in eco-weird refers most prominently to *oikos* as home. More than an identification of nature with the Weird, we identify the eco-weird as a toolbox for analyzing the gap between the familiarity of reality and what

B. H. Onishi (✉)
Penn State Altoona, Altoona, PA, USA
e-mail: byo5042@psu.edu

N. M. Bell
Dallas College, Dallas, TX, USA
e-mail: nbell@dallascollege.edu

B. H. Onishi, N. M. Bell (eds.), *The Call of the Eco-Weird in Fiction, Films, and Games*,
https://doi.org/10.1007/978-3-031-77126-2_1

we find. In his genre study on weird fiction, Michael Cisco identifies the Weird with a departure from the familiar (Cisco 2021, p. 31). As such, we argue that the eco-weird simultaneously articulates the sometimes-subtle gap between home and not home, and the discomfort with the home in which one finds oneself. Home, at least in most environmental contexts, often marks something to do with nature. We can think of the Earth as our *arche* home from which our culture, society, and meaning have grown. But home need not be limited to nature in a traditional sense. As developed in the following chapters, part of the eco-weird is to challenge our understanding of home at all, and to potentially seek a new home beyond our familiar planetary orbit. Such challenges bring the *oikos* at the root of the ecological and the *heimlich* (belonging to the home) at the heart of *unheimlich* (uncanny) to the forefront of our analysis. In our fears of environmental disaster and our hopes for environmental flourishing, we enact a departure from our world that enables the imagination to become a partner in the fecundity and generativity of the world.

Perhaps more than other types of criticism, the eco-weird holds open the possibility of challenging genre conventions and environmental canon. Multiple authors in this volume appeal to Nicole Seymour's concept of "bad environmentalism," which, according to Gry Ulstein in "Forms and Themes of the Eco-Weird," opens up the "self-limiting sanctimoniousness of ecocriticism in productive ways." Both Ulstein and Germaine (in "Tabletop Eco-Weird") articulate the eco-weird within a broad spectrum of play that challenges the critical distance between observer and observed and implicates us in the active generativity of our world. The weird, in this volume, cannot only be a static description of an artistic form, but is also a way of being in the world and of performing the world. But as weird, we contend that this performance is taken up as much by the more-than-human world as it is by the human world. We respond to a general sense of becoming, as entangled, messy participants, and highlight the ever-weirding process of our reality. As such, and following Dan O'hara's claim in "A Doxa of the American Weird" that the weird is "a process of becoming" (O'Hara 2021, p. 25), this volume shows that it is partially through our participation in play that we identify the "weird" in environmental analysis.

By design, we do not offer a definitive formulation of "the weird." Instead, following Roger Luckhurst, we argue that the weird is more like "an inflection or tone" (1045) that refuses to "fit narrative or generic expectation" (1050) and "twists or veers away from familiar frames and binary distributions" (Luckhurst 2017, p. 1052). We take up our environmental crisis as a project that necessarily must turn away from the status quo operation and, as Cisco articulates, offer a "departure from familiarity as such" (Cisco 2021, p. 31). We also follow in the footsteps of work done by Mark Fisher, who describes the weird as "a sensation of *wrongness*" so that "a weird entity or object is so strange that it makes us feel that it should not exist, or at least it should not exist here" (Fisher p. 15). Fisher goes on to argue that because the weird entity or object is present, we must reorganize our mechanisms of making meaning. That is: "The weird thing is not wrong…it is our conceptions that must be inadequate" (Fisher p. 15). The eco-weird works to reformulate those conceptual tools to better make sense of ongoing environmental weirding. Along with Elizabeth Parker, we thus contend that the eco-weird is not merely a genre of fiction, film, and games. Rather, it is a kind of "flavoured mode" of criticism (Parker 2020, p. 33). Or, like Merleau-Ponty, who claims that phenomenology is a "style of thinking" (Merleau-Ponty 2006, p. iix), the eco-weird is a kind of style of criticism that encompasses a multivalent and multidisciplinary approach that includes ecocriticism, game studies, eco-phenomenology, environmental hermeneutics, and speculative philosophies. We take up the project of the environment as weird and work to both reveal the weird and to encourage our weird departure.

Perhaps the most definitive thing about the eco-weird is the manifestation of new worlds via an extrusion of a reality that is already existent. In many instances of the weird, especially in prominent versions of the "Old Weird"[1] established by authors like H.P. Lovecraft and Arthur Machen, there is a focus on an intrusion from some outside force. It is the break with reality and the introduction of a new world that causes characters to go mad and the world to be revealed anew. The eco-weird, while open to intrusion, is more focused on the weirdness of our material world. It takes what has been known about our reality and shows it to be weirder than previously thought. That is, it weirds the world as given, whether it is desert dunes, kitchen tables, fungal gardens, or extra-orbital space. In a kind of looping turn of Cisco's claim that the weird is "a departure from

[1] There is not a clearly technical definition between what is considered Old Weird and what is considered New Weird. However, in broad terms, Old Weird refers to the cosmic horror of H.P. Lovecraft, Algernon Blackwood, and Arthur Machen. It is also closely associated with pulp magazines. The New Weird arises in response to the Old Weird's conservative politics and embraces a largely inclusive stance on bodies, gender fluidity, and social construction. For more, see Noys, Benjamin, and Timothy S. Murphy 2016. "Introduction: Old and New Weird." *Genre* 49 (2): 117–134.

familiarity as such," we argue that the departure is from a world of projection that covers over the already-present weird. The project of the eco-weird, then, is to uncover and to reveal the weird reality already lurking beneath the surface.

This kind of extrusion can be seen prominently in the problem of climate change. There is clearly horror in such changes and destructive patterns, which is why Christy Tidwell and Carter Soles look explicitly to ecohorror as a critical lens. According to Tidwell and Soles, "Ecohorror reflects this estrangement and reveals the horror of *knowing* we live on a terraformed planet, one not terraformed for our benefit." This leads them to claim that "ecohorror may be the dominant mode in which we talk to ourselves about the global climate crisis and the real-life ecological horrors of our current Anthropocenic moment" (Tidwell and Soles 2021, p. 3). While we agree that there is horror in the confrontation of our climate changed world, we also argue that we cannot be limited in our analytical approach to fear or anxiety. Instead, we weird the analytical playing field and address the world not only in fear but also in awe and wonder.

A compelling aspect of horror is that it deals with the world we are presently living in. More than science fiction (which is forward looking) or fantasy (which is otherworldly), horror is about our world right now.[2] In both horror and in the Old Weird, there is a focus on an ontological breach that shakes up our world at a foundational level.[3] While subtle, the eco-weird pushes against the necessity of intrusion. Claims about global weirding have been generated by rising temperatures, changing weather patterns, a simultaneous increase of flooding and droughts, and so on; the conditions for these weird patterns are embedded within the known laws of physics, biology, and climatology. That is, no intrusion from some outside force is generating this weird weather. Instead, it is the result of dangerous anthropogenic activity that runs directly into the mechanisms of a working system. It is as though we depart our world only to finally be returned to it. Therefore, the eco-weird, at least in part, works to depart the familiarity of our world, to generate a strange "sensation of wrongness" at the heart of our home, and to offer tools for analyzing an ever-weirding world.

As a genre and a critical category, the weird is already a transgressive mode of analysis; it makes the familiar unfamiliar and therefore provides an opportunity to chart a new path. This is true for both analytical approaches

[2] See Cisco (2021, p. 27).
[3] See Carroll (1990, p. 16).

and for activities of meaning making. According to Steven Shapiro, the weird "highlights how contemporary formations of knowledge production are themselves barriers to comprehending the past" (Shapiro 2021, p. 64). While this sense of disruption and awareness-raising is similar in other modes of analysis such as the Gothic, Shapiro goes on to claim that "the weird has a greater self-awareness about the creation of dominant epistemology and the ideological barriers to the structuration of knowledge that created an initial sense of frightful bewilderment in the first place" (Shapiro 2021, p. 64). Shapiro's articulation provides support for the claim that what we are doing here is weirding analysis as such, and that the weird is a mode that is particularly good at poking at the status quo. Johnny Murray echoes some of Shapiro's understanding when he says that "the weird takes certain estranging elements essential to its constituent genres and gives them each a further twist, so to speak, estranging the very estrangements in such a way that their generic associations are thoroughly contaminated with one another and subsumed into something new—something *weird*" (Murray 2021, p. 37). While Murray is speaking explicitly about genre, something that he has in common with Cisco, he emphasizes the entangled nature of the weird as an approach.

Eileen A. Joy also highlights the transgressive possibilities present in the weird but focuses explicitly on the act of speculative reading. For Joy, "a speculative reading practice might pay more attention to the ways in which any given unit of a text has its own propensities and relations that might pull against the system and open it to productive errancy" (Joy 2013, p. 29). In this practice, how we read a text has the power to make and remake our reception of that text. According to Joy, "making things (such as a novel, or a poem) that are weird even more weird is … an ethical act, one invested in maximizing the sensual and other richness of the world's expressivity" (Joy 2013, p. 30). Joy's description of speculative reading relates to the intersection of the weird and environmental hermeneutics offered in this volume by Nathan M. Bell. For Bell, the eco-weird attends to the gap between human and nonhuman. This is especially pronounced in the act of translation because the nonhuman is never fully understood by the human, and vice versa. Bell takes Joy's idea that attending to the weird as a text, in this case nature or the nonhuman other, can allow us to see the world differently. The invocation of weird fiction makes this connection explicit. The use of fiction dramatizes and allows the weird of the nonhuman world to stand out. It is the enactment of the weird, which, as Jon Newell says, "creates aesthetic encounters which help us to think about the unthinkable" (Newell 2020, p. 4).Or, as Joy puts it, the weird can "allow texts a certain anti-reductionist and autopoetic, yet also intermediate, ontology" (Joy 2013, p. 32). The knots of the world untangle to reveal something that is both unexplainable and, yet, fecund and generative.

Ultimately, the eco-weird highlights the conjunction of stories and analysis because, as Elizabeth Parker contests, stories about the environment provide an opportunity to "collectively [work] through what we think and feel about the nonhuman environment" (Parker 2020, p. 4). While the impassable gap remains between observer and observed, between the critic and their text, the eco-weird charts a path deeply into these stories so that we are fundamentally entwined with the generative worlds and narrative arcs. Through the eco-weird, we thus emerge back into an extruded world.

As noted above, there is no singular definition of the eco-weird. The eco-weird may be thought of as a subgenre of fiction, as a method of engaging with fiction, or as a conceptual tool for rethinking the human–nature relationship, among others. In Chap. 2, "Fungal Fictions: New Weird Materialism and Mycelial Biohorror," Jonathan Newell considers the eco-weird through fungal horror in New Weird fiction. Both Gothic and older Weird fiction emerged as responses to the disenchantments of the ages in which they emerged. Likewise, the New Weird has emerged in response to the age of ecological crisis, when we are aware of both our destruction of the natural world and our entanglement in it. Considering various fiction examples, Newell shows how the eco-weird is not about the nonhuman world as a horrific other, but about the unavoidable entanglement of the human and more-than-human. The eco-weird is, Newell contends, as much about awe as it is about horror, and it points to the potential of a collaborative future where we have moved past human mastery over the natural world.

In Chap. 3, "Departing the Place Once Familiar: Lovecraft's Eco-Weird Thought," Brian Hisao Onishi considers the importance of place in the work and thought of Lovecraft. Onishi argues that place is integral to both the environmental thought and the racism apparent in Lovecraft's work. Drawing on both speculative realism and eco-phenomenology, he considers the eco-weird as a tool for rereading and revealing the ecological thought found in Lovecraft's Weird fiction. The eco-weird allows us to recontextualize Lovecraft's thought, both going deeper into the ecological relevance of place and challenging the racism that plagues Lovecraft's work.

In Chap. 4, "The Weird as Crisis Discipline: Ontological Tipping Points in Algernon Blackwood's 'Sand'," Emily Alder considers the aforementioned story as an example of eco-weird fiction, particularly highlighting the ecological themes found in Weird stories. Blackwood's "Sand," Alder argues, presents the desert, particularly the Eastern Sahara, as an entity with its own agency. The main character's encounter with the desert shows the desert to be a distinct type of life that goes beyond normal human modes of understanding, which thus requires that we consider various ways of knowing and engaging with the natural world. Similarly to Newell, Alder shows how the eco-weird as a genre invites us to consider both our destruction of the natural world and our reciprocity with it.

In Chap. 5, "Weird Ecology and the Deconstruction of the Globe," Brad Tabas uses weird ecological thinking to reconsider our usual ideas of spatial relations. Tabas considers the idea of tracking as a way to reevaluate our narrow views on mapping and to think through the complexities of spatial relations. He further considers deconstruction and pragmatism as ways to rethink the usual boundaries between the globe and space, between Earth and Heaven, and between what is inside and outside. Like Onishi, Tabas focuses on work by Lovecraft, in this case considering how Lovecraft's fiction helps us to think about exo-spatiality and how we can disrupt the usual boundaries between what is considered earthly and alien. Tabas argues we need a new type of weird ecological thinking, which includes tracking, as a way to rethink our usual relations.

In Chap. 6, "Hermeneutics and the Eco-Weird," Nathan M. Bell considers how the eco-weird and environmental hermeneutics can overlap in considering the insurmountable otherness of the natural world. Bell argues that hermeneutics and translation lead to the understanding that more-than-human others can never be fully known or understood. For Bell, hermeneutics and the eco-weird both encourage us to consider the plurality of meanings, consider different modes of understanding and engaging with more-than-human others, and ultimately both lead us to consider the radical alterity of the natural world. A weird environmental hermeneutics, Bell argues, shows that we need to avoid forcing the natural world into our concepts and accept the otherness of more-than-human others.

In Chap. 7, "(Eco)-Weirding Folk Horror in Alex Garland's *Men*," Dawn Keetley attends to the eco-weird in film. As suggested in her title, Keetley argues that the film *Men* rethinks, or weirds, the usual conventions and tropes of folk horror. In particular, the film pushes both the urban/rural dichotomy and the image of the green man to consider not just the human-nature relationship but to consider class division and more particularly gender division through the trauma of the main character. The film also challenges the usual anthropocentric gaze prominent in folk horror.

In Chap. 8, “Staying with the Weird: Apophatic Wonder and Cosmographic Exploration in Eco-Weird Games,” Nicholas J. Mizer examines the eco-weird through table-top roleplaying games (TRPGs). As with other chapters in this volume, Mizer considers the way that the eco-weird expands beyond the usual theme of Weird horror, considering the relationship between both horror and awe. In particular, TRPGs engage the imagination and world-building of the players, cultivating a sense of wonder about the potentialities of the future in relation to the weirdness, and horror, of the ecological crisis of our world.

In Chap. 9, “Tabletop Eco-Weird: Gameplay Experience and Ecological Ethics,” Chloé Germaine likewise considers the eco-weird through TRPGs. Germaine uses the eco-weird as a type of “aesthetic experience” that helps us rethink our relationship with the natural world. Utilizing multiples examples she argues that tabletop games are inherently weird, even when they do not have the themes or images that we typically associate with Weird fiction. Games are weird precisely because they involve multiple people co-creating worlds, along with the pieces and rules of the game itself, that force us to consider the entanglements that make up the environment and reconsider how we relate to and dwell within the natural world.

In the final chapter of this volume, “Forms and Themes of the Eco-Weird: Experimentation and Play in a Warming World,” Gry Ulstein considers how the eco-weird includes not just the themes of Weird fiction but the type of reading involved. Many Weird texts do not just contain ecological themes but engage readers more deeply through various techniques that challenge the reader. Using various examples, Ulstein argues for the potential of experimental narratives to draw the reader into a type of play in which they are co-creators of the story, which helps readers to think more critically about our entanglement with the more-than-human world.

It is no accident that the various chapters of this volume use “eco-weird” in slightly different ways as there is no neat, simple definition of the term. And yet, throughout this volume we can see several consistent themes. Whether through fiction, film, or games, the eco-weird involves rethinking the relationship between the self and the (ecological) other, highlighting and foregrounding not only the other but also our messy, entangled coexistence with the more-than-human world. The eco-weird disrupts our usual view of the self and the other, and shows us how the environment that is our home is weird and strange, in ways that can be both horrifying and wonderful. Likewise, there is no neat outcome for eco-weird analysis. While there is hope in the awe and wonder that such analysis often entails, there is also horror and unease at the current state of affairs. We offer the eco-weird as a tool for analysis, for engaging in a myriad of textual styles and formats, but we do not prescribe the construction of what comes after our departure from the familiar.

References

Carroll, Noël. 1990. *The Philosophy of Horror or Paradoxes of the Heart.* New York, NY: Routledge.

Cisco, Michael. 2021. *Weird Fiction: A Genre Study.* Cham, Switzerland: Palgrave Macmillan.

Fisher, Mark. 2016. *The Weird and the Eerie.* London: Repeater Books.

Joy, Eileen A. 2013. Weird Reading. *Speculations* IV: 28–34.

Luckhurst, Roger. 2017. The Weird: A Dis/orientation. *Textual Practice*: 1041–1061.

Merleau-Ponty, Maurice. 2006. *Phenomenology of Perception.* Translated by Colin Smith. New York: Routledge Classics.

Murray, Johnny. 2021. The Oozy Set: Toward a Weird(ed) Taxonomy. In *The American Weird: Concept and Medium*, ed. Julius Greve and Florian Zappe, 28–39. New York, NY: Bloomsbury Academic.

Newell, Jonathan. 2020. *A Century of Weird Fiction, 1832–1937: Disgust, Metaphysics and the Aesthetics of Cosmic Horror.* Cardiff: University of Wales Press.

Noys, Benjamin, and Timothy S. Murphy. 2016. Introduction: Old and New Weird. *Genre* 49 (2): 117–134.

O'Hara, Dan. 2021. A Doxa of the American Weird. In *The American Weird: Concept and Medium*, ed. Julius Greve and Florian Zappe, 15–27. New York, NY: Bloomsbury Academic.

Parker, Elizabeth. 2020. *The Forest and the EcoGothic: The Deep Dark Woods in the Popular Imagination.* Cham, Switzerland: Palgrave Macmillan.

Shapiro, Stephen. 2021. Woke Weird and the Cultural Politics of Camp Transformation. In *The American Weird: Concept and Medium*, ed. Julius Greve and Florian Zappe, 55–71. New York: Bloomsbury Academic.

Tidwell, Christy, and Carter Soles. 2021. Introduction: Ecohorror in the Anthropocene. In *Fear and Nature: Ecohorror Studies in the Anthropocene*, ed. Christy Tidwell and Carter Soles, 1–20. State College: The Pennsylvania State University Press.

CHAPTER 2

Fungal Fictions: New Weird Materialism and Mycelial Biohorror

Jonathan Newell

2.1 The Time of the Mushroom

The Gothic emerged in part as a reaction to the crises and contradictions of the Enlightenment, a tenebrous reminder that the supposedly rational, rapidly industrializing world was indelibly stained with blood. The Weird, erupting a century later, registered a similar, arguably even more profound shock responding to the desacralization of the universe in the wake of advancing scientific knowledge, the spread of global capitalism, and the violent horror of imperial decay. Both modes can be understood as literary responses to the traumatic convulsions of modernity and their attendant disenchantments, the way that empire, technoscience, and ideology were transforming the human subject and its relation to reality. The Gothic gives us uncomfortable self-knowledge, exploring truths about ourselves and our history, that which the eighteenth-century cult of reason sought to repress or transcend; the Weird confronts us with the ineffable

J. Newell (✉)
Langara College, Vancouver, BC, Canada
e-mail: jnewell@langara.ca

B. H. Onishi, N. M. Bell (eds.), *The Call of the Eco-Weird in Fiction, Films, and Games*,
https://doi.org/10.1007/978-3-031-77126-2_2

strangeness of the nonhuman, as we became increasingly aware of our own relative unimportance in a vast and meaningless cosmos.

In his article exploring both the Gothic and the Weird, New Weird author China Miéville suggests that the former is best exemplified by the skull, "the outstanding synecdochic signifier for a revenant human dead … mind-seat now empty-eyed, memento mori, grinning, screaming" (2008, p. 124), evoking a sense of "a present stained with traces of the ghostly, the dead-but-unquiet" (p. 112). The skull, Miéville argues, resonates with the Gothic's inherently hauntological and uncanny preoccupations, its fixation on spectrality and the return of the repressed. Gothic texts are exhumations, not just of corpses but of family secrets, hidden crimes, histories that have been buried and willfully forgotten. As Fred Botting observes, it is no coincidence that the Gothic emerged at a moment "of political, economic, and social upheaval," in which the self-consciously secular, scientific, rational, progressive ideology of late eighteenth-century Europe invented a shadow for itself, "a past of chivalry, violence, magical beings, and malevolent aristocrats … bound up with the shifts from feudal to commercial practices in which notions of property, government, and society were undergoing massive transformations," which in turn produced reassessments of our "ideas about nature, art and subjectivity" (2012, pp. 13–14). Even as the Enlightenment sought to contrast itself with a barbaric medieval past, Gothic fiction frequently metaphorized the crises not of the past but of the then-present—the industrial revolution, the emergence of capitalism, the violent extension of empire, bourgeois political apotheosis. Horace Walpole's *The Castle of Otranto* (1764) is, of course, a mythic reimagining of the feudal past, but it is also saturated with its author's Whig sympathies, as critics like Crystal Lake have noted (2014, p. 491); the Gothic novels of Anne Radcliffe, while sometimes misread as "escapist romance" (Watt 2014, p. 67), are centrally concerned with issues of class and gender at a time when both were radically changing; Mary Shelley's *Frankenstein* (1818) incarnates in the patchwork body of its monster a host of anxieties around race, class, and animality which cannot be divorced from the tumult and catastrophe of its era. The Gothic, in other words, shows us the skull beneath the skin: a past we cannot forget, a future we are still bloodily fashioning, parts of ourselves from which we cannot ever escape.

As Miéville is quick to note, however, despite superficial similarities, the Weird "is starkly opposed to the hauntological" (2008, p. 112) and to the entire aesthetic strategy of the Gothic: "the Weird is not the return of any repressed" (p. 113), less the uncanny than what he calls the "abcanny," a "radical otherness" (2010, p. 380) characterized by an aggressive "beyond meaning-ness" (p. 382). As H.P. Lovecraft himself famously puts it with characteristic grandiloquence in *Supernatural Horror in Literature* (1927):

> The true weird tale has something more than secret murder, bloody bones, or a sheeted form clanking chains according to rule [...] a certain atmosphere of breathless and unexplainable dread of outer, unknown forces must be present; and there must be a hint, expressed with a seriousness and portentousness becoming its subject, of that most terrible conception of the human brain – a malign and particular suspension or defeat of those fixed laws of Nature which are our only safeguard against the assaults of chaos and the daemons of unplumbed space. (2011, p. 27)

For Miéville, the Weird is best symbolized not by the human skull but by a distinctly nonhuman body part: the tentacle, "a limb-type with no Gothic or traditional precedents (in 'Western' aesthetics)" (2008, 1p. 05). He ties this shift to a changing understanding of reality and our place within it, arguing that the Weird evokes "a chaotic, amoral, anthropoperipheral universe" (Miéville 2008, p. 112), one revealed by the scientific revelations of the nineteenth century and by the horrors of the First World War. Of course, it would be a mistake to associate the "anti-meaningness" of the Weird with apoliticism. As modern disillusionment eroded what remained of Enlightenment optimism, the Weird—especially in its Lovecraftian versions—conjured a vision of decline, decay, and degeneration, one all-too-often intertwined with the reactionary movements of the early twentieth century. Much Weird fiction of this time reveals a deep disgust with the alien, particularly nonhuman beings already among us, contaminating, polluting: the malign "Little People" of Arthur Machen's "The Novel of the Black Seal" (1895), the abominable swine-men of William Hope Hodgson's *The House on the Borderland* (1908), the amphibious Deep Ones in Lovecraft's "The Shadow Over Innsmouth" (1936). The tentacle is frequently a violating and sinister appendage. For all that older Weird fiction might reveal ontological dimensions previously obscured by anthropocentrism, it is also inclined to forms of what Simon Estok and others have called "ecophobia," "a generalized fear or contempt for the natural world and its inhabitants" (2011, p. 4), a dread of our inevitable entanglement with the nonhuman.

If the late eighteenth and early nineteenth centuries were the Time of the Skull, when writers of the Gothic gazed into the darkness of human history and human nature whose banishment the Enlightenment vaingloriously attempted, and the late nineteenth and early twentieth centuries inaugurated the Time of the Tentacle, when writers of the Weird recoiled in revulsion from a universe which suddenly seemed ancient, unfathomable, and hideously indifferent, then we have now entered a new epoch. Our conceptions, both of ourselves and of the nonhuman world, have once again dramatically shifted as we become more and more cognizant not only of our entanglement with the "natural" world around us but also of our capacity to poison, corrupt, and destroy the tenuous ecological networks in which we are inextricably enmeshed. It is fitting, then, that writers of what has been called the New Weird or eco-weird have seized on a new motif, a recurrent trope that runs rampant through works of the late twentieth and early twenty-first centuries, one which suddenly seems to be everywhere, sprouting, spreading, sporulating. It is now, I will argue, the Time of the Mushroom.

This chapter focuses on the recurrent trope of monstrous fungi in recent Weird fiction, suggesting that the relentless profusion of fictional mycelia constitutes a New Weird novum, one especially fecund with ecological meaning and which serves as a cultural diffraction of emerging posthuman subjectivities and an accompanying sprawl of environmental politics: fungal feminisms, mycelial Marxisms, psychedelic postcolonialisms. As Orrin Grey and Silvia Moreno-Garcia observe in their recent collection of fungal stories, "there's a small but persistent thread that runs, mycelium-like, through the history of weird fiction" (2012, p. 7). While Weird fungus first germinated in the early twentieth century with William Hope Hodgson's "The Voice in the Night" (1907), it has recently propagated through a range of texts and other media, from literature to video games to films and television. I begin with a brief consideration of Hodgson's sporous story to tease out some of the differences between the tentacular and fungal Weird, or Weird and eco-weird, suggesting that "The Voice in the Night" foreshadows more recent texts. I then trace several tendrils of the mycelial Weird through the works of authors like Jeff VanderMeer, Aliya Whiteley, and Silvia Moreno-Garcia, drawing on new materialism to read Weird fungi as incarnating the anxieties of environmental decay and, simultaneously, the possibilities of a posthuman future. Just as a great deal of earlier Weird fiction does not literally involve tentacular monsters, so does much of the eco-weird eschew the fungal; my aim here is to illuminate, by the bioluminescent glow of the works in question, some of the distinctive features of recent eco-weird fiction in contrast with their literary forbears and to think through what such fungal fictions might tell us about our moment in history and our entanglement with the nonhuman world. I also hope to show how eco-weird fiction is especially well-equipped to address the representational problems engendered by climate change and the Anthropocene. As Dawn Keetley argues, narratives about mold "are distinctively able to tell stories about species death in a warming world" (2021, p. 45); indeed, I want to suggest that fungal fictions constitute a kind of literary equivalent to the philosophical and theoretical approaches of the "nonhuman turn," such as new materialism and speculative realism. Fungal fictions offer a vision of apocalypse, not only in the sense of the end of the world, but in the original sense of the term—as a disclosure or revelation, a vision from the cosmic Outside. The fungal eco-weird can be read as a cultural omen, a symptom which might also be a savior, expressing dread and longing both at once: fungus comes not only to destroy but also to remake the world in a mycelial messianism.

2.2 Lichenous Posthumanism in "The Voice in the Night"

Hodgson's "The Voice in the Night" has been largely neglected by most scholars, but as Emily Alder has recently suggested, the story exemplifies how the Weird can explore "the implications for humans of sharing an entirely physical, embodied existence with other organisms" (2020, p. 159), to show how "boundaries collapse across kingdoms of matter ... and are reconstructed into new, interstitial forms" (p. 188). In the story, a shipwrecked couple find themselves castaway on a deserted island infested with a sinister lichen that, they discover, has already engulfed and assimilated the crew of a previous vessel. Despite their efforts to keep themselves free from contamination, the couple gradually succumbs to a creeping fungal rot, even ingesting the fungus in a scene which darkly parallels the original sin of Genesis. The story is narrated by the husband of the pair, John, who pleads with the crew of another vessel for food while keeping out of the light to avoid revealing the extent of his bodily transformation—until, in the final sentences, a stray beam of dawn sunlight illuminates something like "a great, grey nodding sponge," a "Thing" that quickly disappears into the morning fog (Hodgson 2018, p. 161). The creature is a version of what Kelly Hurley famously calls the "abhuman" (a term she in fact borrows from Hodgson)—"a not-quite-human subject, characterized by its morphic variability, continually in danger of becoming not-itself, becoming other" (1996, pp. 3–4). As the narrator describes himself and his wife, the "fungoid growth" expands "day by day, with monstrous rapidity" until the pair effectively cease to be human: "Nothing we could do would check it materially and so – and so – we who had been human became – Well, it matters less each day" (Hodgson 2018, p. 160). Hurley herself, in her brief treatment of the story, observes that the transformation of the narrator into a "Thing" constitutes the abhuman *par excellence*, in which "the demarcations between 'the human' and any number of not-human configurations are inexorably erased" (1996, p. 31). The fungus is more than a debilitating or disfiguring infection, but exerts a kind of unsettling, parasitic agency; the narrator notes that "desire for it had entered into our blood," producing a terrible "hunger-lust" (Hodgson 2018, p. 160). In this sense, the fungus undermines faith in the supremacy of human reason and autonomy and, with it, narratives that place human beings at the top of natural hierarchies, instead suggesting that neither our bodies nor our minds are stable or impermeable, but rather always vulnerable to the tendrils of the nonhuman.

Through its display of what Hurley would term "the ruination of the human subject" (1996, p. 2), "The Voice in the Night" seems a very typical Weird tale, superficially comparable to the works of Lovecraft and his ilk. We might compare the fungal contamination that overtakes its protagonists to similar preternatural horrors, such as the nameless "colour out of space" from Lovecraft's story of the same name, an alien radiation that infects all it touches. Much like the narrator of "The Voice in the Night," the characters in "The Colour Out of Space" (1927), including a married couple, are slowly consumed by a seeping, inexorable doom. Nahum Gardner confines his wife to the attic as she deteriorates, eventually killing her, while he himself succumbs to "collapse, greying, and disintegration," with "horrible brittleness" and "dry fragments … scaling off" (Lovecraft 2014b, p. 329). In a scene strikingly similar to the reveal of the nodding sponge at the end of "The Voice in the Night," the narrator of "The Colour Out of Space" finds Nahum unrecognizable, looking "horrifiedly into the distorted parody that had been a face" (Lovecraft 2014b, p. 329); much like the lichenous speaker in Hodgson's story, Nahum also becomes thinglike, given the pronoun "it" as he completely caves in (2014b, p. 330). In the stories of Lovecraft—and, indeed, many of Hodgson's own stories, and those of other Weird authors—the porosity and dissolution of the subject in the face of the invasion of the nonhuman is apprehended with profound ecophobia. The apprehension of the agency of the nonhuman and its permeation of the world around us and within us is accompanied by visceral disgust, loathing, and above all a sense of ***antagonism***—a feeling that human civilization and human beings ourselves are fragile, like to regress and decay back into primeval squalor. We might think also of the dissolution of Helen Vaughan at the end of Arthur Machen's *The Great God Pan* (1890), a slippage down the evolutionary ladder into "the abyss of all being" till all that remains is a kind of unclean "jelly" and finally "a horrible and unspeakable shape" (2018, p. 50) or of the bestial ghosts in the stories of M.R. James, animalistic spirits with "coarse black hairs, longer than ever grew on human hand" (2015, p. 474) bringing to mind apes and gigantic spiders, clawing at the borders of an ever-precarious social order, products of fin de siècle fever dreams, imperial paranoia, and the trauma of post-Victorian future shock.

While "The Voice in the Night" is clearly not free of this ecophobia, the fungal story foreshadows recent reinventions of the Weird. Where those of Lovecraft's protagonists contaminated by the horror of the nonhuman are typically reduced to gibbering madness if they manage to avoid dissolving entirely into cosmic putrescence like the Gardners, one of the chief victims of "The Voice in the Night" not only survives contact with the nonhuman but eloquently tells the story of his own extensive transformation. Previous castaways, though less articulate, seem to have merged with the lichen, becoming part of the island, spared from what otherwise seems certain starvation. The lichen is hardly framed as the savior of the couple, and certainly its description still emphasizes its horror—it quakes "vilely," and rises in "horrible, fantastic mounds" (Hodgson 2018, p. 157) implied to be other humans it has assimilated—and yet, in the couple's becoming-fungus, ecophobia mingles with something like what Timothy Morton calls "ecognosis"—a process of "becoming accustomed to something strange" in such a way that the "strangeness doesn't become less strange through acclimation" (2016, p. 5). The story's end is not hopeful as such—but neither is it catastrophic. As John rows "nodding into the mist" the distinction between "hand and oar" (Hodgson 2018, p. 161)—between body, boat, lichen—gives way to what new materialists like Jane Bennett would term an "assemblage," a figure of "thing-power" beyond the "life-matter binary" (2010, p. 20). The fungus in this story not only infects but also feeds; for all the horror of John and his fiancée, the lichen provides nourishment, a means of survival. As Bennett points out, food can be understood not as inert matter but as "conative bodies vying alongside and within another complex body (a person's 'own' body)," an "actant inside and alongside intention-forming, morality-(dis)obeying, language-using, reflexivity wielding, and culture-making human beings" (2010, p. 39). In becoming part of John, his fiancée, and the rest of the castaways, the fungus makes viscerally visible the everyday reality of all food. The familiar is revealed as strange all along; reality is indelibly weirded. John has become what Morton would call a "strange stranger"—something which "is itself and uncannily not itself at the same time" (2016, p. 18). But the fungal does not register ecognosis merely with the neurotic pessimism of an Arkham academic faced with the incursions of alien beings, nor with the suicidal courage of Gustaf Johansen ramming Cthulhu in the emptiness of the southern Pacific. John's fate may be imagined, but it is never fully described; he rows into an uncertain, posthuman future.

2.3 Sporous Ecognosis in the Ambergris Cycle

From the beginning, then, the fungal bears the germ of something radical. Though there are certainly fungal fictions throughout the twentieth century—it crops up in a pastiche of Hodgson's story called "The Fungus Isle" (1923) by Phillip Fisher, in various stories by Stephen King such as the memorably revolting "Gray Matter" (1973), in Harry Adam Knight's recently reprinted *The Fungus* (1983)—the fungal has since spread throughout the pages of the New Weird, and nowhere more spectacularly than in the works of Jeff VanderMeer. A number of scholars have considered VanderMeer's work in relation to ecology and to older Weird fiction. Gry Ulstein suggests that VanderMeer's fiction seeks to "move beyond cosmic fear" (2017, p. 75) but engages with the monstrous and otherworldly more ambivalently, asking "whether it is really 'such a bad thing' to be colonized, assimilated, altered, and forcefully evolved" (2017, p. 92) by nonhuman actants. I want to build on Ulstein's argument, paying particular attention to VanderMeer's signature motif of fungal contamination, which Ulstein mentions only briefly.

I am also particularly indebted to Alison Sperling's extensive work on VanderMeer, especially her contention that in works like *The Southern Reach Trilogy*, VanderMeer presents us with "a particularly 'weird' ecology" or "body-ecology," one that "illustrates the sick body and its ecological surroundings as the site of an active and ongoing set of relations" (2016, p. 232). Drawing on ecofeminism and disability studies, Sperling suggests that these texts are a form of "ecosickness fiction" exposing "a deep wound made by the intrusion of the Anthropocene into the body, a feverish corporeality" (2016, p. 233). Following Sperling's claims, I read VanderMeer's fungus as producing a form of ecognosis, one that specifically comments on the destructive consequences of anthropocentric thinking. Rather than focusing primarily on the *Southern Reach* books, however, I turn my attention here to VanderMeer's earlier work, what is sometimes called the Ambergris cycle: the short story collection *City of Saints and Madmen* (2002), the fantastical biography *Shriek: An Afterword*, and the Weird noir novel *Finch* (2009).

Ambergris—named after the "most valuable and secret part of the whale" (VanderMeer, 2006a, p. 7)—is a city in a secondary world, but, unlike the feudalism of most post-Tolkienian fantasy, a world undergoing the spasms and transformations of modernity. A center for commerce, art, music, and especially publishing, Ambergris is founded by a ruthless whaler, pirate, and colonizer, Cappan John Manzikert, later styled Manzikert I. Stories in the Ambergris cycle range in time period from the city's founding through to a period which, by the time of *Finch*, vaguely resembles the early twentieth century, complete with automobiles, typewriters, telephones, dirigibles, and tanks. The city is also thoroughly, ineradicably infested by fungus, spreading from the subterranean world that lies below its streets: all that remains of Cinsorium, the fungal city of the gray caps or mushroom dwellers which predates Ambergris, and who are viciously displaced and driven underground in the city's early history. Across the three books, VanderMeer slowly reveals the nature of the gray caps, a species of interplanetary or perhaps interdimensional refugees struggling to return to their home who eventually rise up against Ambergris and seize the city in an anti-colonial mycelial reoccupation. Throughout the Ambergris cycle, I suggest, fungi and the gray caps come to represent elements of the nonhuman world which settler-colonial capitalism exploits, marginalizes, and represses in ways that ultimately engender its own self-destruction. Simultaneously, fungus also produces forms of posthuman metamorphosis and becoming, creating the possibility of a more mutualistic and sustainable future in contrast with a fundamentally doomed and deluded anthropocentrism.

VanderMeer's work subtly but consistently repudiates what we might, following Karen Barad, call the "ethic-onto-epistemology" (2007, p. 90) of the Old Weird, and the ethical, epistemological, and ontological entanglements it depicts; in the Ambergris cycle, this dynamic manifests with particular salience through the fungal. In *The Hoegbotton Guide to the Early History of the City of Ambergris*, one of the novellas within *City of Saints and Madmen*, the narrator, the historian Duncan Shriek—who becomes the protagonist of the next book in the cycle—frames Cinsorium as a city of wonders, describing "the high domed buildings in sparkling good repair, the amphitheaters looking as they might, within minutes, play host to innumerable entertainment" (VanderMeer 2006a, p. 115). Cinsorium, Duncan suggests, may have been "among the largest of all ancient cities" (VanderMeer 2006a, p. 114) and shows all evidence of "a healthy culture" (p. 115), with artworks including an elaborate "mechanized golden tree" (p. 119) and "mosaics … made from lichen and fungus skillfully placed and trained to achieve the desired effect" (VanderMeer 2006a, p. 118). Such are the splendors of the place that Manzikert and his followers, including the monk Samuel Tonsure from whose journals Duncan is working, are awed—even as, simultaneously, they express disgust and disdain at the sight of the gray caps themselves, who Tonsure declares a degenerate species living in "undeniable decadence" (VanderMeer 2006a, p. 115), language which clearly evokes eugenicist anxieties of imperial decline. For evidence of this supposed regression, Tonsure asserts that the gray caps are "unable to understand the processes of the city, unable to work its machines, unable to farm" (VanderMeer 2006a, p. 115). He describes a library, the shelves of which "had rotted away before the onslaught of a startling profusion of small dark purple mushrooms with white stems," the books "made from palm frond pulp" spilling "upon the main floor—thousands of books, many ornately engraved with strange letters and overlaid by the now familiar golden lichen" which suffuses and beautifies much of Cinsorium; Tonsure and Manzikert watch in horror and contempt as the gray caps burn the books "as firewood" (VanderMeer 2006a, p. 116). The gray caps themselves Tonsure describes as "runts" and "cripples" who "had become unfit to rule" the city "or even live within its boundaries," with semi-fungal bodies becoming pale and sickly, "mocking the memory of what they had been" (VanderMeer 2006a, p. 119).

Tonsure's description evokes similar passages in Lovecraft's novella *At the Mountains of Madness* (1936), in which Antarctic explorers discover a vast, alien city of unfathomable age, seemingly abandoned. Lovecraft's story displays the same sense of awe at the "incredible, inhuman massiveness" of the city, the way it stretches "for boundless miles in every direction" (2014a, p. 507)—but also, simultaneously, the casual, almost laughably chauvinistic judgment of the city's denizens. William Dyer, the narrator, deems some of the carvings of the city "mature, accomplished, and aesthetically evolved to the highest degree of civilized mastery" yet simultaneously "utterly alien in every detail" (Lovecraft 2014a, p. 519), but when he and his fellow explorers reach another part of the city later confirmed to be newer than the previous, he asserts the carvings are distinctly "decadent in comparison" (p. 523), the product of "slackened energies and aspirations of a stressful and uncertain period" (p. 538). Again and again, a "certain decadence" (Lovecraft 2014a, p. 540) is emphasized as the Old Ones of the city succumb to climatic disaster and the uprisings of their protoplasmic slaves, the amorphous shoggoths. All of this, of course, is strongly resonant with Lovecraft's own terror at the prospect of "Western decline," a theme which infamously recurs throughout his fiction. As many have remarked, stories like "Pickman's Model" (1927), "The Horror at Red Hook" (1927), and "The Shadow over Innsmouth" (1936) explicitly link this decline to white supremacist narratives of racial decay. Much as Tonsure's consideration of Cinsorium's architecture echoes *At the Mountains of Madness*, so does his description of the gray caps recall such figures as the Deep Ones and their half-human progeny suffering from the "Innsmouth look" (Lovecraft 2014c, p. 590), a condition compared to a "cutaneous disease" (p. 584) which the narrator speculates might result from "biological degeneration" or some admixture of "foreign blood" (p. 585), a clue to their partially piscine heritage.

In VanderMeer's case, however, the assessments of the explorers are revealed as ignorance and bigotry. Duncan suggests that Tonsure's interpretation of the library and its book-burnings is woefully inaccurate, arguing that the library "is actually a place for religious worship, and the 'books' prayer rolls" (VanderMeer 2006a, p. 116) which are ritually burnt. He points out that the seemingly decaying shelves "were rotted wooden planks *specifically* inserted into the walls to foster the growth of the special purple mushrooms, which grew nowhere else in Cinsorium and may well have had religious significance," a supposition reinforced by the presence of a mushroom-shaped altar outside the structure (VanderMeer 2006a, pp. 116–117). He observes that the "lawns covered in mushrooms of every possible size and color – from gigantic growths as large as the Oliphaunt to delicate, glassy nodules no longer than a child's fingernail" which produce "the most varied and remarkable fragrances" (VanderMeer

2006a, p. 112)—clearly constitute a form of agriculture, contrary to Tonsure's insistence that the gray caps were "reduced to hunting and gathering" (p. 115). The very fungi which Tonsure and Manzikert read as signs of decay are in fact marks of the gray caps' sophistication, a sign that they—unlike the brutal colonizers—understand themselves as interconnected with other organisms and living sustainably alongside them. Notably, it is a mushroom which Manzikert takes as justification for his genocide of the gray caps, after discovering "a small, flesh-colored lichen" bearing "a striking resemblance to the Lepress Saint Kristina of Malfour" being menaced by "a lichen that look uncannily like a grey cap" (VanderMeer 2006a, p. 121). Interpreting this sign as a holy call to arms, he leads a slaughter of the gray caps which culminates in the utter destruction of Cinsorium's surface in preparation for the founding of Ambergris. Manzikert and Tonsure, in other words, view the fungi and the gray caps as Lovecraftian protagonists might—as the symptoms of decadence and barbarism and social rot, a society growing more "primitive" or closer to nature, and as confirmation of their own racial and cultural superiority. Through Duncan Shriek's commentary, however, VanderMeer turns these assumptions on their head, showing that it is the colonists' own ultimately fatal misreading of the fungus which, ironically, proves precisely the opposite, revealing the explorers as imperial brutes incapable of fathoming the complex civilization into which they have wandered.

The two sequels to *City of Saints and Madmen*, *Shriek: An Afterword* and *Finch*, both depict a series of crises that ultimately spell Ambergris's doom and metamorphosis. Through the characters of Duncan Shriek—the same historian from *City of Saints and Madmen*—and the hardboiled detective John Finch, we see not only the city's gradual unraveling as the gray caps rise up to reclaim their occupied territory but also the transformation of the Ambergrisian settlers in the wake of this uprising. *Shriek* is framed as an afterword penned by Duncan's sister Janice to *The Hoegbotton Guide to the Early History of the City of Ambergris*, but quickly takes the form of a meandering biography of both siblings, annotated by Duncan himself. The book reveals the truth of the mysterious "Silence," the chilling disappearance of many of Ambergris's citizens early in its history, a disappearance previously understood as an act of revenge on the part of the gray caps but which, Duncan learns, was actually the result of a malfunction from a subterranean "Machine" designed to return the gray caps home. As Duncan says to Janice, "the Machine is a door … but the flaw in it wasn't about the door itself. It was the location. They have to bring it aboveground. They have to reclaim the city" (VanderMeer 2006b, p. 323). *Finch* depicts this undertaking, as the gray caps emerge from their chthonic burrows and occupy Ambergris, their "monstrous fungal cathedrals [rising] under cover of darkness, replacing the old, familiar architecture" (VanderMeer 2006b, p. 13). Morton suggests that "Ecological awareness is that moment at which ... narrators find out that they are the tragic criminal" (9); it is precisely this kind of awareness that gradually comes into focus, the gray caps revealed as far more than Lovecraftian grotesques, the narrators, in coming to better understand the fungal creatures and the crimes of their own forebears, finding themselves transformed in the process.

As in *City of Saints and Madmen*, fungus runs riot in both *Shriek* and *Finch*, appearing at times as an antagonistic force, a contaminating nonhuman actant, but ultimately becoming the medium for ecognosis and posthuman becoming. As Anna Tsing writes in her book *The Mushroom at the End of the World* (2015), which considers how nature manages to live despite the environmental destruction wrought by capitalism, "collaboration means working across difference, which leads to contamination"—but without collaboration "we all die" (p. 28). The fungal in the Ambergris cycle is never some all-benevolent force of Nature sweeping away the evils of human civilization, nor is it a crude alien abomination to be ripped out and destroyed. In *Shriek*, Duncan, increasingly obsessed with the gray caps and the Silence, repeatedly delves into the undercity in search of truths hidden beneath the physical and ideological surface of Ambergris. He returns with glimpses of repressed and cryptic knowledge, but he is also changed, his body made a host to fungi from below. Arriving at Janice's door after one such expedition, he is "covered in mushrooms and other fungi," organisms growing into his body: "the fungi had needled his head, burrowed into the skin, forming whorls of brightness that hummed with fecundity" (VanderMeer 2006b, p. 33). From his eyelashes to his fingernails, he is covered in spores. As Octavia Cade observes, Duncan's very biology has become just as "transgressive" as his historical investigations (2021, p. 117) in ways that dramatize how human bodies are never purely human but are themselves ecosystems, holobionts combining a host body and various mutualistic organisms (p. 108). Duncan eventually transforms into something even stranger, his body wracked with bouts of fungal disease which he experiences simultaneously as ecosickness and as a kind of rapturous longing, a "yearning" in his flesh which "calls out to the yearning in the ground" (VanderMeer 2006b, p. 164), "spores infiltrating [his] head, [his] limbs," "clambering over [his] sinuses," getting "between [him] and [his] own skin," seething across his eyes to bring "a stinging green veil over [his] sight" (p. 165). What Duncan's increasingly hybrid body makes visible is what Stacy Alaimo terms "trans-corporeality," the idea that "all creatures, all embodied beings, are intermeshed with the dynamic, material world, which crosses through them, transforms them, and is transformed by them" (2018, p. 435). His scholarship earns him the scorn of publishers, his colleagues, and eventually his lover and later academic rival, Mary Sabon, but increasingly Duncan becomes a figure both uncanny *and* abcanny; all that Ambergris has forgotten and deliberately repressed, he brings to light.

By the time of *Finch*, set a century later than *Shriek*, Duncan's fungal symbiosis has spread to the rest of Ambergris. The novel takes the form of a hardboiled detective story, another genre, like the Weird, premised on a quest for knowledge; as Morton says, "the darkness of ecological awareness is the darkness of noir" (2016, p. 9). By the time of the novel, some of the city's citizens have become "Partials," human collaborators with the gray caps which smell like "sweet rotting meat" (VanderMeer 2009, p. 4), their skin covered in "tiny fruiting bodies" (p. 3), their eyes becoming strange fungal cameras in what Finch calls a "permanent occupation" (p. 12). But even for those who refuse this Faustian bargain, fungus is unavoidable, replacing technology and food, filling the air with spores, drugging citizens, and watching them with organic surveillance technology. The novel, told through Finch's viewpoint, at times presents the gray caps, much as Lovecraft might, as invaders from below despoiling Ambergris, corrupting its citizens. Figures like Wyte, for instance, one of Finch's fellow detectives who becomes infected with spores, are consumed from the inside out, rendered utterly unrecognizable by fungal rot. Wyte's body becomes "*other*," we are told, "a garden of fungus," his "arms ballooning out into sudden wings of brilliant purple-red-orange," his voice "incomprehensible and strange," "guttural and animal-like"; in the end, "no part of him … was human" (VanderMeer 2009, p. 184). Wyte's body is "monstrous," "spilling out in peculiar ways," his subjectivity "dissolving" into an abhuman mass, a "*thing with Wyte's eyes*" (VanderMeer 2009, p. 252). "There's not much of me left" (VanderMeer 2009, p. 252), Wyte tells Finch, as he loses control of his own body, the fungus taking over as he loses his pronoun, becomes an "it" which smiles horrifically even as Finch and several other men dismember him, his body persisting "with the reflexive life of a gecko's tail" (p. 255) but with "no relation to the family man and husband Wyte had once been" (p. 256). In such moments, the fungal seems to invite an ecophobic reaction, something only to be opposed and expunged to restore human civilization.

As the novel progresses, however, Finch's relation to the fungal becomes more complicated. The man whose murder he has been investigating is revealed as none other than Duncan Shriek himself, his life extended by mycelial symbiosis. After Finch eats the "memory bulb" that grows from Duncan's body—a standard investigative technique in occupied Ambergris—Duncan manifests in Finch's head as a kind of fungal ghost. At first, Finch recoils, calling Duncan "a fucking parasite" (VanderMeer 2009, p. 298), but Duncan insists he is rather "a passenger" (p. 299), suggesting that he and Finch can "open" to one another, allowing Duncan to temporarily assume control of Finch, healing him of his injuries while simultaneously borrowing from his tissues in order for Duncan to construct a new body from himself, eventually passing through an eldritch gate in order to marshal forces to save the city. The novel ends not with some neat resolution, the gray caps banished forever, the fungal excised and Ambergris cleansed, but with an uneasy moment of reflection as Finch—in an eerie scene reminiscent of the end of "The Voice in the Night"—rows along the river and "watches the end and beginning of history" (VanderMeer 2009, p. 334) as rebels summoned by Duncan begin overrunning gray cap positions.

Throughout the Ambergris cycle, the fungal is both horrific and wondrous at once. It can and does corrupt and consume, but it also transforms, heals, and enlightens. It is this fundamental ambivalence, I think, which distinguishes the fungal eco-weird from the tentacular Weird of old. A nonhuman actant caught in a mesh of politics, languages, cultures, and other human and nonhuman actants, it transits across bodies and minds, sometimes colonizing and controlling, sometimes forming webs of interconnection. At times it reflects our own capacity for violence, a biopolitical and thanatopolitical technology, molded by its users into weapons, systems of surveillance, narcotizing methods of pacification. At others, it permits forms of resistance and rewilding, reclaiming that which was despoiled. In some moments, especially in *Finch*, it seems to queasily occupy both positions at once, just as it is at once a part and not a part of those to whom it attaches and infests. Always changing, never static, it destroys and creates, disgusts and sometimes delights, reforms and revolts.

2.4 Mycelial Motherhood in "The Beauty"

More recent eco-weird novels have deployed the fungal even more hopefully than the Ambergris cycle. In Aliya Whitely's novella *The Beauty* (2014), what at first seems a tale of ecophobic horror and the monstrous feminine slowly transforms into something not only far stranger but also far more optimistic. The story takes place in a postapocalyptic future decimated by a fungal epidemic implied to have killed everyone with a uterus. Set in the Valley of the Rocks in England, the story revolves around the remnants of a utopian commune, the "Group." Having lost its female members to the fungus, a thick, yellow infection which creeps out from bodily orifices, the Group now consists only of men, who lack any means of continuing the species. As the protagonist, the Group's storyteller, Nate, declares, "There will be no humanity after us" (Whitely 2018, p. 9). Over the course of the text, however, mysterious yellow mushrooms sprout from the graves of the Group's women and eventually reveal themselves as uncanny fungal doubles of dead women which Nate collectively dubs the "Beauty." While searching for two missing men, Nate stumbles into one of these creatures, a being "yellow and spongy and limbed, with a smooth round ball for a head" but distinctly feminine, with "breasts, globes of yellow, and rounded hips that speak to [Nate] of woman, of want, and that [disgust him] beyond words" (Whitely 2018, p. 20). This initial revulsion soon gives way to a lust, however, as Nate relents to the advances of the fungal doppelganger he calls "Bee," succumbing to the seduction of her "mother hum" (Whitely 2018, p. 18) and "cold yellow arms" (p. 23), joining with her physically and telepathically. Eventually, Nate returns to the Group with more of the Beauty, who became part of the social fabric despite the protestations of some of the Group's members. Even as violence against the Beauty escalates, the Beauty impregnate their lovers with hybrid fungal-human children and transform their bodies, minds, and social norms to rear their progeny together. The novella climaxes with the murder of several of the Beauty and the violent abortion of two unborn hybrid children, after which the Beauty kill most of the Group. In the wake of this violent attempt to reassert patriarchal control and the consequent massacre, those left of the Beauty without a mate become the hyper-aggressive "Unloved," held back from assaulting the few remaining men only out of fear of harming hybrids, born or gestating. Having directed Bee to overthrow the domineering would-be patriarch Ted, his own uncle, and throw him to the Unloved, the now-pregnant

Nate resolves to leave the Valley of the Rocks with Bee to find a better future for himself and his child.

Initially, the Beauty are clearly described through the language of the abject. When Nate first meets Bee, he is dropped into a wet, muddy hole filled with the resonant "mother-sound" of the Beauty from which he "cannot run" and which very explicitly brings "back memories of something like Mother" (Whitely 2018, p. 18)—very clearly a uterine space. He experiences Bee's body as contaminating and disgusting, so powerfully so that upon first meeting her he vomits and soils himself (Whitely 2018, p. 20). Eventually, though, his revulsion mingles with a queasy desire to touch Bee; when he does, he discovers that he "can't separate its desires from [his] own," as if the two had become one being. When they have sex, he feels that they "are drawn into [themselves], completely without the world," and he begins to experience visions "of earthy darkness, of waiting, growing, moving to sunlight, opening, learning and expanding," a process explicitly compared to "being a baby in a womb, deep in the mother and unaware of anything but that sharp, tingling and delicious edge of potential" (Whitely 2018, p. 24). The scene plays out as a reversal of what Julia Kristeva terms "abjection," the psychic process of cleaving from the mother's body to establish a sense of individual subjectivity: Nate is swallowed up by the mother in a kind of reverse birth, his individuality subsumed, reincorporated into the maternal whole. To encounter the abject, Kristeva writes, is to be drawn along a "twisted braid of affects and thoughts" toward "a vortex of summons and repulsion" (1982, p. 1), back into a state of queasy, pre-Oedipal oneness, back into the Real. When, at the end of Part One, Nate leads the Beauty back to the Group and they stalk into the circle of firelight, his story ending, they "encircle" the men, "enclose them, take them into their embrace and rock them, absorb them, until terror and pleasure become one and the same," and narrative gives ways to "sighs and sobs of wordless confusion" (Whitely 2018, p. 29). It is as if the patriarchal logos of the Group and the Symbolic Order they maintain has been upended; the Beauty are silent, communicating entirely through images, prelinguistic, primevally corporeal.

If the story had ended there, *The Beauty* might read as a quasi-Lovecraftian short story, the titular fungal women an instance of the ravenous Outside destroying the fragile civilization of men, monstrous and feminized others akin to various "degenerate" ghouls, Deep Ones, and similar horrors whose nonhumanness, in the original Weird, frequently threatens to subsume white masculinity. Here, though, something more complex unfolds. As the Beauty integrate themselves into the Group and eventually impregnate Thomas, their cook, several things become clear. Firstly, the Group's leaders are deeply anxious about the Beauty—both the political threat they clearly pose to their own authority and the deeper psychological threat they pose to the masculinity of the Group. Uncle Ted tells Nate quite frankly that "there are those among us who want to tread [the Beauty] back down into the mushrooms they sprang from," warning him against tipping the group "into chaos" (Whitely 2018, p. 56). What Ted leaves out is what point there is to order without any hope for a future; as Nate reproaches, "maybe we want to live on with the Beauty. Maybe we don't want your protection" (Whitely 2018, p. 57). Ted's response is tellingly gendered: "You're not more than a man after all. You're less than one. You always will be" (Whitely 57). The escalating efforts of Ted and others to maintain their control are not grounded in any pragmatic consideration for self-preservation or even the continuation of a particular culture or society: without the Beauty, all human (or, at least, humanoid) life and civilization must be presumed doomed. Rather, the custodians of the Symbolic Order are desperate to maintain boundaries between masculinity and femininity and between humanity and the nonhuman purely for their own sake: they would rather commit slow suicide than relinquish their power. As the Beauty and their lovers seek to produce a queer, posthuman future, however, this thanatotic death-wish deepens into ecocidal testeria.

The obsessive boundary-policing of the Group's leaders manifests as revulsion and hatred. "It disgusts me," William says of the Beauty—"It should disgust us all" (Whitely 2018, p. 46). The efforts of the patriarchs to excise the Beauty read like the actions of Old Weird protagonists, the enactment of what Kristeva would term a "ritual" to "ward off the subject's fear of his very own identity sinking irretrievably into the mother" (1982, p. 64). Here, misogyny mingles with ecophobia; what the elders of the Group fear, in notable contrast with the more accepting adolescents, is not merely femininity or nature as such, but the possibility that men might allow themselves to be feminized or penetrated. In other words, they cannot acknowledge the borders of the (masculine) self as porous, cannot understand the human body's entanglement in an ecological network in anything but horrific terms. Filled with "contempt and fear … for the agency of the natural environment" (2011, p. 5), as Estok describes ecophobia, the Group's reigning council refuse to recognize what Nate, in his stories to the Group after the Beauty's arrival, begins to apprehend:

that the only way forward for humanity is to become something other than merely human. The end of *The Beauty* bears some similarity to both "The Voice in the Night" and *Finch*, as Nate and Bee walk "a path that unravels one step at a time, on and on, with no end" (Whitely 2018, p. 103). Proclaiming that he is "sick of words," that he is setting out "not even to find new stories to tell" (Whitely 2018, p. 103), Nate sloughs off the patriarchal Symbolic Order and the world of language to embrace queer motherhood and a world of psychic images, becoming one with the nonhuman Beauty and the child growing inside him. This, too, is ecognosis: Nate declares they are seeking "what remains of humanity ... Just for the sake of knowing" (Whitely 2018, p. 103), conjuring possibilities of a fungal, xenofeminist future rejecting gender binaries of male and female along with ecophobic binaries of human and nonhuman.

2.5 Rhizomatic Resistance in *Mexican Gothic*

Silvia Moreno-Garcia's novel *Mexican Gothic* (2020), even more than the work of VanderMeer and Whiteley, directly responds to the Lovecraftian Weird and imperial Gothic. Moreno-Garcia has a background in science studies and literary criticism; her master's thesis, *Magna Mater: Women and Eugenic Thought in the Work of H.P. Lovecraft* (2016), shows how Lovecraft's fiction is steeped in a misogynistic and racist anxiety centered around the female body. As she puts it, "order in Lovecraft's fiction is precarious and white men are in constant danger of being infected or destroyed" by dangerous "Others" (Moreno-Garcia 2016, p. 7). Her own novel deliberately subverts these very tropes, using a preternatural fungus known as "the gloom" to illustrate how anthropocentrism and the various structures of oppression it undergirds cannot sustain their own contradictions. The text initially seems a classic tale of what some might call the Female Gothic: the protagonist, the socialite Noemí Taboada, journeys to the decrepit mansion of High Place in the mountains of Mexico, hoping to help her ailing cousin Catalina after receiving a cryptic and sinister letter suggesting that her husband, Virgil Doyle, is poisoning her. As Noemí investigates this dark possibility, she chafes against the strict English customs of the Doyle house and the openly eugenicist philosophy of Howard Doyle, the decrepit patriarch of the inbred Doyle family, a figure named after Lovecraft and Arthur Conan Doyle. As time passes, Noemí begins to experience bizarre nightmares and uncovers something of the Doyles's troubled family history, one stained by murder and exploitation, including the death of Ruth, Howard's niece, who had fallen in love with an indigenous miner and who shot Howard before killing herself. In time, Noemí discovers the truth: Howard Doyle is far older than he seems. He is using the gloom—a fungal organism he stole from a community who once used it for medicine, and which now infests High Place—as a kind of psychic network, one which can encode human consciousness and memories inside itself. Having sacrificed the body and mind of his wife, Agnes, to act as a hub for the mycelial network of the gloom, he has been transmigrating his mind into the bodies of his offspring in a Saturnian cycle, father devouring son. To her horror, Noemí also learns of her own intended place within this monstrous plot—as the Doyle bloodline has exhausted itself, they now seek fresh genetic material to rejuvenate their failing dynasty, and wish to marry her to Francis, whose body Howard intends to possess. Rather than become another sacrifice, Noemí joins Catalina and Francis and burns High Place to the ground, including both Agnes and Howard. The gloom, however, survives in Francis (and, perhaps, Noemí herself), who is stayed from suicide to end the familial "curse" by Noemí's reassurances.

In the first half of *Mexican Gothic*, the gloom appears as a singularly sinister force, strongly associated with High Place itself. Waking, Noemí glimpses it as a mold afflicting the wallpaper of the house; sleeping, its tendrils manifest in her dreams, sprouting and spreading to become nightmares of the dead, consciousnesses trapped within the mycelial network in a fashion similar to the psychic projections of Duncan Shriek in VanderMeer's *Finch*. Initially, these hauntings seem like classic ghosts—"a woman in a dress of antique lace," for instance, with "a glow, golden like the mushrooms on the wall" where her face should be (Moreno-Garcia 2020, p. 56). In a clear allusion to Charlotte Perkins Gillman's "The Yellow Wallpaper" (1892), Noemí wonders whether there is something "strange lurking behind the wallpaper" (Moreno-Garcia 2020, p. 57) as she stirs from sleep. High Place serves as a symbol of colonial oppression, "a little piece of England" (Moreno-Garcia 2020, p. 18) transported to the Mexican landscape; as the gloom and High Place blend together, the mold of the gloom encrusting mirrors and furniture, becoming teeming forests of mushrooms in Noemí's dreams, fungal infection becomes a colonization, infiltrating Noemí's mind and body. At times, it seems to override her agency, filling her with unwelcome and alien impulses, including a "sweet, sickening desire" that makes her "roll her hips, sinuous, a serpent" (Moreno-Garcia 2020, p. 81), the Doyle's heraldic animal, as she dreams of Virgil Doyle, eldest scion of the family line, sometimes mingled with visions of his diseased and elderly father. She begins to sleepwalk, controlled by the fungus, her agency eaten away as surely as mold worries the wallpaper. As Noemí finds herself trapped in High Place, she glimpses the fate of Agnes, sacrificed to the gloom:

> The fungus would erupt up, from her body, up through the soil, weaving itself into the walls, extending into the foundations of the building. And the gloom needed a mind. It needed her. The gloom was alive. It was alive in more than one way; at its rotten core there was the corpse of a woman, her limbs twisted, her hair brittle against the skull. And the corpse stretched its jaws open, screaming inside the earth, and from her dried lips emerged the pale mushroom. (Moreno-Garcia 2020, p. 218)

The gloom is Howard Doyle's weapon, his means of extending his life and his mastery, of maintaining his position at the top of the social hierarchy; through it, eventually, he can attempt to manipulate Noemí like a puppet, compelling her to move at his command. "The house loves you," Doyle insists, "*You're ours, like it or not. You're ours and you're us*" (Moreno-Garcia 2020, p. 219). Seen in this light, the fungal serves as a technology of power, a means of discipline and control—not the horror of some alien Real outside of the Symbolic Order, but the ultimate manifestation of that order. Howard's fungus-ridden body with its crusting of "large, dark boils … as thick as barnacles upon a ship's hull" (Moreno-Garcia 2020, p. 203) is as disgusting as any Lovecraftian monstrosity, but here it is whiteness and patriarchal masculinity rendered monstrous rather than miscegenation or femininity.

As *Mexican Gothic* progresses, however, and the gloom's origins are revealed, it emerges as more than an extension of empire but as a force of interconnection, solidarity, and knowledge. Francis observes that fungi "can enter into symbiotic relationships" (Moreno-Garcia 2020, p. 211), and though Howard has exploited this ability to transfer his consciousness from body to body, Noemí discovers that the gloom also provides *her* with access to its repository of mycelial memory. The visions she experiences are revealed not merely as Howard's malignant attempts to assert control over her body, to reduce her to a breeding vessel, but as the whisperings of his previous victims conveyed through the gloom: "A memory? A ghost? Not quite a ghost. She realized that what she had been seeing, the voice whispering to her, urging her to open her eyes, was the mind of

Ruth, which still nestled in the gloom, in the crevices of the mold-covered walls" (Moreno-Garcia 2020, p. 233). While Howard used the gloom to force Ruth to shoot herself, in turn what remains of Ruth uses the gloom to warn and help liberate Noemí. We are told that in its original form the gloom was used to heal wounds and preserve health, worshipped as a god in a cave-shrine with fungus "hanging from the cave walls" instead of candles (Moreno-Garcia 2020, p. 205). As the full extent of the Doyles's crimes becomes clear—including the deaths of numerous Mexican miners, exposed to the gloom but unaccustomed to it—the gloom becomes multivalent, representing not only colonial oppression but *also* those that empire exploits and consumes. While Howard uses the gloom in a transhumanist bid to attempt to overcome nature, to become a Cartesian mind free from the flesh, this is not its only use; it also serves as a means of contesting colonial and patriarchal power, one that might look beyond the Anthropocene and Capitalocene. Though the novel ends with High Place, Agnes, and the gloom burned, the fungus lingers in the blood of Francis, and possibly Noemí as well. As Noemí ponders, "she wondered whether one day, if she looked carefully, she might notice a golden sheen" to Francis's eyes, a sign of the gloom's infection—"or maybe she'd catch her own reflection staring back at her with eyes of molten gold" (Moreno-Garcia 2020, pp. 300–301). This thought, however uneasy, intertwines with another—the "hope that the world could be remade, kinder and sweeter" (Moreno-Garcia 2020, p. 301). The novel resists definitive resolution, just as *Finch* and *The Beauty* or even "The Voice in the Night" do, but as in these earlier texts, the fungal is more than some Lovecraftian doom, an alien horror gnashing at the borders of unplumbed space; it is a nonhuman actant inextricably entangled with us in a mesh of filaments, an unfinished tapestry which we and other beings are continually weaving and reweaving.

2.6 Becoming-Fungus

In the Time of the Mushroom, an era wracked by climate disasters, dread of the nonhuman world and its terrifying power is to be expected. The eco-weird, however, frequently attempts to sidestep and subvert the familiar and often reactionary ecophobia common in older Weird fiction, which registers the human body's porosity and the enmeshment of our social systems in ecological networks with panic and disgust. On the one hand, fungus appears in these texts at times as a potential weapon wielded by the powerful against the marginalized, inverting older Weird dynamics that enmonstered the feminine, the colonized, the nonhuman. The fungal fictions I have briefly discussed here—just a sampling of many such sporous texts currently proliferating throughout our culture—do not deprive the nonhuman of its capacity for terror or even revulsion. But, consistently, loathing and hatred of the nonhuman is revealed in eco-weird stories as patriarchal desperation, imperial testeria, the hubris of agricultural logistics, a delusional and ultimately suicidal anthropocentrism incapable of understanding the omnipresent agency and intra-action of the nonhuman which inevitably and mutualistically suffuses and entangles us. Fungal fiction frequently enacts a cathartic, violent retribution against ecocidal systems of power, providing productive fantasies of rewilding and regenerative destruction. Such fantasies are rarely a reversion to some idyllic status quo that modernity has disrupted but to a postapocalyptic and posthuman future, inviting eschatological longing, solidarity with the sporous. Eco-weird apocalypse can involve a mystic surrender to the nonhuman world—an ecstatic becoming-fungus, an embrace of the abhuman. If we are to survive, it often seems to suggest, we must relinquish delusions of mastery or purity. The fungal, in its very ambiguity, its capacity to erode boundaries between organisms, self and other, human and nonhuman, is never *just* a contaminating force; rather, it reveals the collaborative potential already-inherent in contamination. Refusing equally blithe utopianism and fatalism, fungal fiction instead suggests a horizon of possibility, a posthuman futurity toward which we must, with bodies teeming and corrupted, row.

References

Alaimo, Stacy. 2018. Trans-corporeality. In *Posthuman Glossary*, ed. Rosi Braidotti and Maria Hlavajova, 435–438. London: Bloomsbury Academic.

Alder, Emily. 2020. *Weird Fiction and Science at the Fin de Siècle*. London: Palgrave.

Barad, Karen. 2007. *Meeting the Universe Halfway. Quantum Physics and the Entanglement of Matter and Meaning*. Durham: Duke UP.

Bennett, Jane. 2010. *Vibrant Matter: A Political Ecology of Things*. Durham: Duke UP.

Botting, Fred. 2012. In Gothic Darkly: Heterotopia, History, Culture. In *A New Companion to the Gothic*, ed. David Punter, 11–24. Oxford: Blackwell.

Cade, Octavia. 2021. Microbiology and Microcosms: Ecosystem and the Body in *Shriek: An Afterword*. In *Surreal Entanglements: Essays on Jeff VanderMeer's Fiction*, ed. Louise Economides and Laura Shackelford, 107–123. New York: Routledge.

Estok, Simon. 2011. *Ecocriticism and Shakespeare: Reading Ecophobia*. New York: Palgrave.

Grey, Orrin, and Silvia Moreno-Garcia. 2012. Introduction. In *Fungi*, ed. Orrin Grey and Silvia Moreno-Garcia, 7–8. Vancouver: Innsmouth Free Press.

Hodgson, William. 2018. The Voice in the Night. In *The Ghost Pirates and Other Revenants of the Sea*, ed. Jeremy Lassen, 150–161. New York: Night Shade Books.

Hurley, Kelly. 1996. *The Gothic Body*. Cambridge: Cambridge University Press.

James, Montague Rhodes. 2015. Canon Alberic's Scrap-Book. In *Classic Tales of Horror*, 467–476. San Diego: Canterbury Classics.

Keetley, Dawn. 2021. Black Mold, White Extinction: *I Am the Pretty Thing That Lives in the House*, *The Haunting of Hill House*, 'Gray Matter,' and H. P. Lovecraft's 'The Shunned House'. In *Haunted Nature*, ed. Sladja Blazan, 43–66. London: Palgrave.

Kristeva, Julia. 1982. *Powers of Horror: An Essay on Abjection*. Translated by Leon S. Roudiez. New York: Columbia University Press.

Lake, Crystal. 2014. Bloody Records: Manuscripts and Politics in *The Castle of Otranto*. *Modern Philology* 110: 489–512.

Lovecraft, Howard Phillips. 2011. *Supernatural Horror in Literature*. Cabin John: Wildside Press.

———. 2014a. At the Mountains of Madness. In *The New Annotated H.P. Lovecraft*, ed. Leslie Klinger, 457–572. New York: Norton.

———. 2014b. The Colour Out of Space. In *The New Annotated H.P. Lovecraft*, ed. Leslie Klinger, 310–342. New York: Norton.

———. 2014c. The Shadow Over Innsmouth. In *The New Annotated H.P. Lovecraft*, ed. Leslie Klinger, 573–642. New York: Norton.

Machen, Arthur. 2018. *The Great God Pan*. Oxford. Oxford University Press.

Miéville, China. 2008. M.R. James and the Quantum Vampire: Weird; Hauntological: Versus and/or and and/or or? *Collapse* 4: 105–128.

———. 2010. On Monsters: Or, Nine or More (Monstrous) Not Cannies. *Journal of the Fantastic in the Arts* 23: 377–392.

Moreno-Garcia, Silvia. 2016. *Magna Mater: Women and Eugenic Thought in the Work of H.P. Lovecraft*. The University of British Columbia. https://doi.org/10.14288/1.0228100. Accessed 8 May 2024.

———. 2020. *Mexican Gothic*. New York: Penguin.

Morton, Timothy. 2016. *Dark Ecology: For a Logic of Future Coexistence*. New York: Columbia University Press.

Sperling, Alison. 2016. Second Skins: A Body-Ecology of Jeff VanderMeer's Southern Reach Trilogy. *Paradoxa* 28: 214–239.

Tsing, Anna. 2015. *The Mushroom at the End of the World: On the Possibility of Life in Capitalist Ruins*. Princeton: Princeton University Press.

Ulstein, Gry. 2017. Brave New Weird: Anthropocene Monsters in Jeff VanderMeer's *The Southern Reach*. *Concentric: Literary and Cultural Studies* 43: 71–96.

VanderMeer, Jeff. 2006a (2002). *City of Saints and Madmen*. New York: Bantam.

———. 2006b. *Shriek: An Afterword*. New York: Tor.

———. 2009. *Finch*. Portland: Underland Press.

Watt, James. 2014. Anne Radcliffe and Politics. In *Anne Radcliffe, Romanticism, and the Gothic*, ed. Dale Townshend and Angela Wright, 67–82. Cambridge: Cambridge University Press.

Whitely, Aliya. 2018. *The Beauty*. London: Titan.

CHAPTER 3

Departing the Place Once Familiar: Lovecraft's Eco-Weird Thought

Brian Hisao Onishi

H.P. Lovecraft is arguably the most influential horror writer in U.S. history. While many may not know his name, his influence stretches, like the tentacles he unknowingly made famous, from Stephen King to Junji Ito. Something that often draws readers to Lovecraft's work is his intimate sense of place. In both his personal writings and his fiction, we see the influence that New England had on him and how closely linked he felt to the area, going so far as to declare "I am Providence," a quote which also seconds as the epigraph on his grave (Klein 1965, pp. xxv–xxvi). While there are few, if any, clear articulations of his environmental thought in either his fiction or his vast archive of personal writings, this sense of place grounds Lovecraft's thought in the environment and connects it to what I call the eco-weird. In this chapter, I will argue that Lovecraft's sense of place is foundational for both his environmental thought and his racist beliefs. Further, because the place of New England is so important in his fiction, I also argue that his racism is a fundamental part of his fiction. Finally, I will argue that the eco-weird offers a set of tools that help to

B. H. Onishi (✉)
Penn State Altoona, Altoona, PA, USA
e-mail: byo5042@psu.edu

B. H. Onishi, N. M. Bell (eds.), *The Call of the Eco-Weird in Fiction, Films, and Games*,
https://doi.org/10.1007/978-3-031-77126-2_3

analyze Lovecraft's sense of place and challenge his racism, ultimately leading to a more nuanced reading of his fiction. To do this, I will appeal to Lovecraft's own letters and short stories, along with commentary and analysis by S.T. Joshi and Gilles Menegaldo to highlight the complexity of Lovecraft's ideas and how they could possibly relate to environmental thought. I will also situate Lovecraft within a framework associated with speculative realism and eco-phenomenology. Doing so will provide a theoretical background that both establishes and pulls on his sense of place, use of nature, and relation to climate change.

3.1 Lovecraft and the Eco-Weird

The eco-weird is an interdisciplinary approach to the connection between environmental thought and Weird fiction,[1] first introduced in my 2020 chapter "Weird Ecologies in The Happening" (B. Onishi 2020). The aim of the eco-weird is to attend to the ways that the world is Weird[2] and to give a critical toolkit for analyzing this weirdness. While I will argue that climate change is perhaps the most extreme form of that weirdness, it shows up in both the mundane and the catastrophic.

For Lovecraft, Weird fiction is characterized by "atmosphere, not action" (Lovecraft 1982, p. 44). The eerie mood of a campfire showcases the creeping speculation of what lies beyond the cone of light generated by its flames. The world shows up as a kind of quasi-agent, requiring that we reconsider how a place changes the way that we think. In part, the eco-weird is about the continuous but sometimes invisible gap between the known world and the world as it is "in-itself." Thus, the eco-weird attempts to amplify the call of the noumenal realm that eludes our access. But more than an epistemological limit, the eco-weird also acknowledges the ways that we, as a human community, show up as a driving force of destruction; how we, like Cthulhu, upset the governing laws of ecosystems and biotic communities. Finally, the eco-weird highlights the entanglement of human communities and nonhuman communities, drawing on the lack of a clear boundary between one and the other, so that it is unclear

[1] Weird fiction here is taken broadly to include fiction, film, video games, tabletop games, and so on.

[2] I capitalize "Weird" when it is used in a technical way so as to distinguish it from a colloquial sense of "weird," meaning strange or out of place. I also capitalize "Weird" when used to demarcate a genre of fiction. Again, this emphasizes the difference between the colloquial use of "weird" and highlights the technical usage.

when the human ends and the nonhuman begins. This is sometimes wonderful and sometimes awful. And sometimes, in the weirdest moments, it is both at the same time.

There are many inroads into the eco-weird. Much work has been done within ecocriticism by Gry Ulstein, Brad Tabas, Dawn Keetley, Chloe Germain, Jonathan Newell, and Alison Sperling that identifies the relationship between the environment and Weird fiction. My own path into the eco-weird is through a strange, diffractive reading of speculative realism and eco-phenomenology.[3] Speculative realism is buoyed by the work of Quentin Meillassoux and Graham Harman, the latter having written *Weird Realism: Lovecraft and Philosophy*, a book about the philosophical connections between Lovecraft's fiction and object-oriented ontologies (Harman 2012). The general position is that while there is a gap between thinking and being, between the knower and the known, that gap allows for speculation. This contrasts with the so-called correlationism of Kantian metaphysics that puts out of bounds any discussion of the noumenal realm. At least since Kant, the problem has been that "we cannot represent the 'in itself' without it becoming 'for us'" (Meillassoux 2008, p. 4). Speculative realism claims that the world is real in such a way that we can distinguish "between those properties of the world which are a function of our relation to it, and those properties of the world as it is 'in itself'" (Meillassoux 2008, p. 3). Meillassoux acknowledges that this seems like an indefensible position given the nature of thought and means of interaction with the world around us yet maintains that it is possible via speculation. I contend that Weird fiction offers a dramatization of how speculation begins to break down the correlational trap. Some force breaks into our realm of knowledge and collapses the boundaries and borders we thought we knew.

Phenomenology, and eco-phenomenology specifically, puts some of this worry about whether speculation is possible out of play. Famously, phenomenology is not worried about whether it presents a realist or an idealist position, which goes against Lovecraft's strong commitment to materialism (taken as a realist philosophy). However, what phenomenology does is offer a philosophical view that is grounded in experience. According to Charles S. Brown and Ted Toadvine, in the turn to the

[3] I provide a more extensive treatment of the relationship between phenomenology and speculative realism in my monograph, *Weird Wonder in Merleau-Ponty, Object-Oriented Ontology, and New Materialism* (2023).

environment as a reservoir of experience, "phenomenology, led by its own momentum, becomes a philosophical ecology, that is, a study of the interrelationship between organism and world in its metaphysical and axiological dimensions" (Brown and Toadvine 2003, pp. xii–xiii). Here the question is not whether we can get outside of our thought to being, but how meaning emerges within a matrix of experiences. In phenomenology, truth is verified by an intersubjectivity that necessarily includes multiple voices and a variety of experiences. Eco-phenomenology allows for some of those voices to be nonhuman, pushing the boundaries of meaning making to nonhuman biotic communities and abiotic communities. The material world becomes a kind of actor in a dialogue, calling me to make meaning alongside it, entangled with it, wrapped up inside a messy play of strange fleshy existence that continuously expands and contracts.

Lovecraft, in many ways, provides a foundational framework for thinking through the eco-weird, not because he clearly articulated his thoughts on Weird fiction and the environment, but because he articulates a version of the Weird that applies generally to reality. It is as if Weird fiction is the genre of reality.[4] Lovecraft's fiction pulls at the alleged tension between speculative realism and eco-phenomenology by highlighting a reality that seems to exceed our grasp while also including the material agency of the world to push us toward what Michael Cisco has called the "deterritorialization of ordinary experience" (Cisco 2021, p. 7). Lovecraft often sounds like a phenomenologist, describing the emergence of meaning from ancient architectures and unreasonable geographies. In an essay titled "An Autobiography: Some Notes on a Nonentity," he states, "the 'punch' of a truly weird tale is simply some violation or transcending of fixed cosmic law—an imaginative escape from palling reality—since phenomena rather than persons are the logical 'heroes'" (Lovecraft 1965, p. 7). Not only is there an escape from "palling reality," which likely occurs via the gap between our experience and our knowledge, but there is a focus on phenomena as major players in the creation of atmosphere. Lovecraft, thus, offers an opportunity to bridge the difference between speculative realism and eco-phenomenology via the framing of Weird fiction.

Ultimately, Lovecraft's essays and fiction put him squarely within the field of the eco-weird. Through his focus on place, he offers an account of the environment that is both strange and wonderful. Within his version of

[4] This is an observation made by Philip Gellatt in a conversation that took place on August 2, 2023.

New England hides a challenge to the taken for granted and the assumed familiarity of the everyday. While not always environmentally oriented, his fiction takes a cosmic turn, which may have been supplied via his reading of Algernon Blackwood. According to Mike Ashley, Blackwood's fiction more explicitly deals with the Weird in nature, and his influence on Lovecraft potentially provides the space for Lovecraft's cosmic turn (Ashley 1988). If this is the case, then Lovecraft's later fiction easily fits within the purview of the eco-weird.

The combination of Lovecraft and Blackwood also paves the way for thinking about our current climatic and environmental crises through the lens of their Weird fiction. It is possible, that is, to think about climate change as a kind of cosmic horror. Here the source of horror flickers back and forth between humanity's suicidal treatment of our environment and the environment's violent and uncaring reaction to our alleged progress. This last piece maps onto Lovecraft's view that humanity is insignificant when situated within the vast cosmos, which leads him to claim that "the interplay of forces which govern climate, behaviour, biological growth and decay, and so on, is too purely universal, cosmic, and eternal a phenomenon to have any relationship to the immediate wishing-phenomena of one minute organic species on our transient and insignificant planet" (Lovecraft, Selected Letters III: 1929–1931 1971, p. 31). We, as humanity, are but a small set of players on a vast and endless field of cosmic action.

3.2 Materialism and a Phenomenology of Place

From both his personal writings and his fiction, it is clear that Lovecraft had a deep affinity for his hometown of Providence, Rhode Island. Many local landmarks show up in his stories, including the Fleur-Des-Lys house on Thomas Street ("The Call of Cthulhu"), the atmosphere of Federal Hill ("The Haunter in the Darkness"), and the John Hay Library ("The Case of Charles Dexter Ward"). But the relationship between Lovecraft and Providence seems to go deeper than just warm feelings for a hometown. Rather, the place of Providence specifically, and New England generally, forms the foundation for his cosmic stories.

Part of his sense of place is built upon his commitment to materialism and manifests in both his rendering of New England as a place of Weird horror and in his understanding of culture, race, religion, and social organization. According to Joshi, Lovecraft subscribed to two basic tenets of materialism: (1) "all entity is material" and (2) "causality is uniform to such a degree that free will is a myth" (Joshi, The Weird Tale 1990, p. 171). This seems, ultimately, to commit Lovecraft to a kind of realism that eliminates the possibility of nonmaterial entities, a position fortified by his outspoken denial of religion of any kind.[5] Famously, Lovecraft eschewed the effectiveness of supernatural elements in his fiction and instead focused on the relation of horror to reality. In a letter to James Ferdinand Morton, Lovecraft affirms this, claiming that "My big kick comes from *taking reality just as it is*...and then permitting my symbolizing faculty to ***build outward*** from the existing facts" (Lovecraft 1971, p. 140, emphasis in the original). In Steven Mariconda's words, Lovecraft was "compelled to create fiction which, by incorporating his imaginings as extensions rather than negations of reality, accommodated both his intellectual and his emotional needs" (Mariconda 1991, p. 188).

While Lovecraft's ideas are rooted in a materialist framework, it often feels very close to phenomenology. This is especially relevant when considering that phenomenology, at least for Dylan Trigg, "draw[s] our attention to the strangeness of things...through attending to the natural world in an unnatural manner." It does this by "bracketing our pregiven understanding of things" which "persist in the world in a synthesis of both familiarity and unfamiliarity" (Trigg 2012, p. 26). This track of phenomenological investigation requires that we deliberately avoid projecting meaning onto the environment. Instead, we must let meaning emerge within a dialogue with the world. For Merleau-Ponty, this requires that we embrace a strange kind of mode of interaction so that our environment merges with our subjectivity such that the dialogue "arranges round the subject a world which speaks to him of himself, and gives his own thoughts their place in the world" (Merleau-Ponty 1962, p. 153). In other words, not only do we arrange the world via our actions, but the world also makes meaning of our actions for us. It situates our intentions, desires, and understandings by drawing us out via a kind of elemental dialogue.

While his allegiance to realism would keep him from being read as a full-fledged phenomenologist, his use of New England as a foundational setting produces similar results to those described by Trigg. Lovecraft's commitment to both materialism and Weird fiction puts him right at this crossroad between the familiar and the unfamiliar. Lovecraft crosses this intersection via his cosmic horror, which renders what is safest as what is most horrifying. The upending of the familiar and the safe leads to what

[5] For more on Lovecraft's atheism, see Lovecraft, H.P., *Against Religion*. Ed by S.T. Joshi.

some of have called the uncanny, which according to Trigg, "is to be understood fundamentally as an effect, a felt experience that disturbs the body, resulting in a departure from the everyday" (Trigg 2012, p. 27). The etymological roots of the uncanny as *unheimlich* (un-homely) also connects us to the *oikos* of the eco-weird. In both instances, the familiarity and comfort of home is disrupted, made weird, and thereby requires us to depart what we once knew as safe.

In line with Trigg's phenomenology, Lovecraft clearly attempts to pull at the veil of known reality to showcase the horror of the world that escapes our knowledge. As stated in the oft-quoted opening to "The Call of Cthulhu",[6] it is the confrontation with the natural world as something that exceeds our understanding that Lovecraft found terrifying. This is clearly seen in a 1931 Letter to August Derleth, in which Lovecraft writes: "It is my contention that real *weirdness* or imaginative liberation depends on the depiction of something *which does not exist*, or which probably does not exist" (Lovecraft, Selected Letters III: 1929–1931 1971, p. 435, emphasis in the original). Through Lovecraft's Weird fiction, we encounter a world that is pure materiality and yet one that escapes our full grasp. The Weirdness generated by these stories is rooted in a sense of place that is both identifiable and destroyed. Once again, we toggle the line between the familiar and the unfamiliar by straddling the difference between the known and the unknown.

In *A Memory of Place: A Phenomenology of the Uncanny*, Trigg articulates a sense of place that treats the material environment as significant in "shaping, defining, and constituting our sense of self and world" (Trigg 2012, p. XVI). He argues for a strange kind of place that is "fundamentally porous" so that "place is never autonomous in its unity, but forever bleeding and seeping into other places, both those of the past and those of the future (Trigg 2012, p. 17). Here the dialogic world of Merleau-Ponty that offers meaning to the investigating phenomenologist seeps into the body schema of the subject so that the subject and the world are deeply and intimately entangled. Given that Lovecraft's exclusionary emphasis on cultural protection led him to decry New York as a dead city, he would likely have objected to Trigg's understanding of place as porous. Yet, he also acknowledges the gap between our experience of a place and the place itself. In another letter to Derleth, he writes: "There is no reason to

[6] The opening sentence reads: "The most merciful thing in the world, I think, is the inability of the human mind to correlate all its contents" (Lovecraft 1999, p. 139).

suppose that any given human impression has any exact correspondence with any external reality. All that creates a favourable presumption, is a high degree of correlation with the impressions of others and the general body of repeatedly demonstrated fact" (Lovecraft 1971, p. 281). Though sometimes frustrating to our scientific pursuits, this gap between what is given in perception and what is actual in reality is generative for expanding our sense of place beyond its temporally situated manifestation. Stories like "The Case of Charles Dexter Ward," "The Call of Cthulhu," and "The Haunter of the Dark" all enliven Providence by pulling at the tension between the given and the ungiven. This is a place that is populated with what David E. Schultz calls an "antimythology" lined with "a sense of the cosmic or otherworldly" (Schultz 1991, p. 213). By focusing on New England, and more specifically Providence, Lovecraft articulates the fecundity of our horror when the familiar becomes strange, and when the landscapes that fill our dreams also inhabit our nightmares.

3.3 Place and Race

Lovecraft's use of place was undoubtedly deliberate. His stories reflect what Mariconda identifies as his "background," a term used by Lovecraft "to describe his cultural heritage in its very broadest dimension" (Mariconda, Lovecraft's Concept of "Background" 1986, p. 5). Not only did this include the geographical and spatial setting in which he lived, but "social customs, attitudes, and institutions" and even bloodline (Mariconda, Lovecraft's Concept of "Background" 1986, p. 5). Implicit here is the idea that Lovecraft's views on race are deeply entangled with his sense of place in New England. Taking Lovecraft's racism into account is important because it supplies an understanding of how place manifests not only for him in his life, but also in his stories. Lovecraft argued that true horror is found in the usurpation of natural laws, which, for him, includes the mixing of races, cultures, and traditions.

This is, perhaps, most strikingly evident during Lovecraft's time in New York, where he spent less than two years of his life from 1924 to 1926. According to Joshi, "Lovecraft had confessed that he found the atmosphere of New York increasingly uncongenial, causing his pen to dry up; he seemed to require the emotional and aesthetic stimulus of his return to Providence to rekindle his creative fires" (Joshi, The Rise and Fall of the Cthulhu Mythos 2008, p. 49). It is only when he returns from New York to his beloved Providence that he is able to write "The Call of Cthulhu," which "was the first of an unprecedented series of novels and tales—amounting to some 150,000 words—that Lovecraft wrote in a period of eight or nine months" (Joshi, The Rise and Fall of the Cthulhu Mythos 2008, p. 49). It is worth noting here that the move back to Providence occurs around the same time that Lovecraft begins to truly appreciate the writing of Algernon Blackwood (Ashley 1988, p. 7). This co-occurrence signals at least two things: First, it establishes Nature (with a capital N) as a place for cosmic horror and is something that he adopts from Blackwood. Second, Nature as found specifically in Providence fills the stories that he writes during this phase of his career. Ultimately, the outburst of creative writing that occurs upon his return to Providence speaks to the way that place-as-environment influences the foundation of his "cosmic turn." That is, Nature and the environment are at the heart of Lovecraft's mythology and Weird world.

New York acts as an explanatory foil for his affinity for Providence as an environmental and value-filled place. According to Menegaldo, Lovecraft began to associate "the modern city with" a moral kind of decadence "corresponding to a certain dehumanization and regression" (Menegaldo 1981, p. 11). This leads Lovecraft to see New York as "a dead city, cut off from its roots, and resembling a huge decomposing corpse" (Menegaldo 1981, p. 11). Supporting Menegaldo's claim, Lovecraft writes in a letter to Derleth in October of 1929 that "New York is no place for a white man to live. The metropolis is a flurried, garish dissonance of aimless speed and magnitude, hybrid and alien to the core, and without historic roots or traditions. It is an emotional and aesthetic island" (Lovecraft 1971, p. 30). On the one hand the ecological language of roots highlights the environmental sense of place at the heart of both his disdain for New York and his love for Providence. On the other hand, this ecological rooting is tied to his cultural ideals such that hybridity and alien otherness mark New York as dangerous and lifeless. The resonance of his sense of place with his sense of cultural traditions (i.e. Aryan bloodlines and Anglo-Saxon heritage) generates a troubling and insidious mechanism for environmental and cultural judgment.

In part, the question at stake here is whether Lovecraft's understanding of place was necessarily tied to his racist ideas. I argue that there is evidence to suggest that this is the case. For one, we most often see Lovecraft describing nature alongside culturally constructed (i.e. human-made) objects. In another letter to Morton, Lovecraft goes so far as to identify himself with the place of Providence, stating: "these waving grasses and towering elms and brook threaded valleys and stone-wall'd farmsteads and white village steeples *are* ME, MYSELF, I, THE CONSCIOUS EGO"! (Lovecraft 1971, p. 31). What is remarkable here is that he not only sees his identity as entwined with the place, but also that the place is equally made up of "waving grasses" and "white village steeples." Returning to his sense of background, Mariconda claims that "his racialist stance … was partly based in his wish to keep his culture-stream—his one last emotional anchor—safe from erosion by an influx of foreign tradition" (Mariconda, Lovecraft's Concept of "Background" 1986, p. 8). The desire to keep a "continuity" with his heritage leads him to appreciate both the environmental and artifactual makeup of Providence. And given that Lovecraft was a staunch materialist, it seems likely that he thought that all matters of taste were determined by the physical traits passed down from generation to generation. If so, he could then argue for a kind of essentialized racism and superiority of taste that is distinctly built into that material lineage.

It is worth questioning whether such an insidious connection between the environment and culture shows up in his fiction or if they are isolated to his personal beliefs. Mariconda provides an initial affirmative response, claiming that "Lovecraft's bleak philosophy, born of his scientific disposition, inspired the cosmic horror of his tales," and "also led him to the concept of relative values, which reinforced the role of his background … as an emotional anchor" (Mariconda, Lovecraft's Concept of "Background" 1986, p. 12). Further, "the terrible alienness of Lovecraft's outside beings and realms is emphasized by their antipodal difference from the narratives' settings" (Mariconda, Lovecraft's Concept of "Background" 1986, p. 12). Following Mariconda, I argue that they are implicitly present in his fiction in a way that challenges our relationship to nature and that can lead to fruitful analysis that highlights the environment in Weird fiction. This analysis can then be used as a criticism of Lovecraft's implicit racism. One piece of evidence that his sense of place and his culturally based racism manifest in his fiction is seen in the difference between the way he concludes stories set in New York and the stories set in Providence. According to Joshi, "It is of interest to note how Willett's complete eradication of Curwen[7] stands in such stark contrast to Malone's obvious failure to eliminate the age-old horror in Red Hook:[8] New York may be the haven of all horror, but Providence must at the end emerge cleansed of an evil taint"

[7] In "The Case of Charles Dexter Ward".

[8] In "The Horror at Red Hook".

(Joshi, I Am Providence—The Life and Times of H.P. Lovecraft: Volume 2 2013, p. 670). New York stands out as both a dead city without roots and a kind of evil garden of horrific and alien monstrosities. Providence, on the other hand, must be cleansed of the evil that would seek to destroy its purity.

The return to Providence from his time in New York solidifies the "alien" and "other" as a significant source of terror for Lovecraft. According to T.E.D. Klein, a notable Weird fiction author who was influenced by Lovecraft, "Providence was one of the few unfailing sources of solace in Lovecraft's life, all the more so during his years of misery and exile in New York, when he had little to sustain him but the vision of an eventual return home" (Klein 1965, p. xlvii). Lovecraft sees the traditional Georgian architecture and heterogenous cultural milieu or Providence as a comfort against the discontinuity and impure alien mixing of New York. Echoing Klein, Mariconda claims that upon "returning home to Providence in April 1926, he experienced an astonishing rejuvenation of his story-writing faculties, and the tale ['The Call of Cthulhu'] was the first in a series of some of his finest works which he began to turn out" (Mariconda, On the Emergence of "Cthulhu" 1987, p. 58). Important here is that Providence seems to have unlocked Lovecraft's most prolific and influential time as a writer.

The foundational importance of place is made explicitly clear when Lovecraft declares that "*places* are everything to me" (Lovecraft 1971, pp. 214–215, emphasis in the original). Menegaldo helps to situate this claim within an articulation of the place of Providence in racial and cultural terms. He does so by highlighting the distinction between the "'historic' parts of Providence" which are "associated with its glorious past, [and] are inhabited by whites of old Anglo-Saxon stock" and the "decadent quarters" which "are mainly in outlying areas and peopled by individuals from eastern Europe or other regions still more 'exotic'" (Menegaldo 1981, p. 12). He notes that Lovecraft thought negatively of these latter places, which "can be seen as a concrete projection of Lovecraft's racist phobias" (Menegaldo 1981, p. 12). Perhaps the most evident connection between his racism and his sense of place come from a letter to August Derleth in 1929. In that letter he states:

> Home is one's ideal setting if one is to develop one's best attributes, and New York is no place for a white man to live. The metropolis is a flurried, garish dissonance of aimless speed and magnitude, hybrid and alien to the

> core, and without historic roots or traditions. It is an emotional and aesthetic island—and a desert island in the long run. A man belongs where he has roots—where the landscape and milieu have some relation to his thoughts and feelings, by virtue of having formed them. (Lovecraft, Selected Letters III: 1929–1931 1971, p. 30)

As mentioned, Lovecraft views New York as a dead city, cut off from its roots and lacking a clear continuity with what he takes to be a sophisticated line of culture, reason, and virtue.

As a foil to a sense of place rooted in White hegemonic culture, we can look to George Yancy's version of place. In *Across Black Spaces*, Yancy provides a critical view of Whiteness that has systematically and systemically devalued non-White communities. That is, a tradition and background of values have been built up around White cultural spaces that have worked to deliberately exclude and devalue non-White participants by setting up Whiteness as "the transcendental norm" (Yancy 2020, p. 2). As evidenced from the scholarship about and writings of Lovecraft, we can see that he clearly participated in building a deliberate set of values based on a transcendental normalization of Whiteness. For Yancy, "whiteness is a form of perverse collective obsessiveness and self-absorption that is structurally anti-Black and that attempts to destroy anything that stands in its colonial path, its explicit and implicit historical mission, and its ontological supremacy" (Yancy 2020, p. 19). Lovecraft is nothing if not obsessed with his cultural background, the geographies that have continued the cultural norms of that background, and even the architectural elements that constitute that background. New York, in his own words, is "no place for a white man" because it does not abide close enough to his hegemonic ideals. It is "no place for a white man" because it does not perform the obsessive anti-Black sentiment that fuels his beloved Providence of the early nineteenth century.

3.4 Phenomenology and Weird Environmental Places

While there are few ways to work around Lovecraft's explicit racism and the role that place plays in constructing his overall set of values, two things allow for a critical rejection and redemption of his sense of place. First is the very ground of the Weird on which he bases much of his fiction and for which he is well known today. Second is a kind of phenomenological reading of Lovecraft's ideas. To the first point, the Weird, as a "deterritorialization" of the familiar, calls into question the ground on which we stand. Cultural backgrounds and social norms seem to fit this category of "familiar," and thus are necessarily called into question by the irruption of the Weird. In a 1930 letter to Derleth, he writes, "With me, the very quality of being cosmically sensitive breeds an exaggerated attachment to the familiar and the immediate—Old Providence, the woods and hills, the ancient ways and thoughts of New England" (Lovecraft 1971, p. 221). There is a push and pull at play between the familiarity of the place and the desire to make it Weird. The place is in danger only because one is deeply rooted in it. The sense of Whiteness and belonging that Lovecraft experienced in Providence is fully on display precisely when it is in danger. This is a contrast to his time in New York where *he* was othered. In New York, the weirdness of place is not the foundation of the Weird, because it is already unfamiliar.

Generating an atmosphere of the eerie, like he does in "The Colour out of Space" when we are told that "West of Arkham the hills rise wild" (Lovecraft 1999, p. 170), highlights the porosity of the place described by Trigg. But this departure arrives alongside "an augmented familiarity" which results in an open tension between the given and the impossible (Trigg 2012, p. 27). The result of Weirding a familiar space that is rooted in a tradition of exclusionary Whiteness is that it potentially disrupts the foundation on which it is built. In this vein, we can read "The Shadow Over Innsmouth" less as a story about the othering and devaluation of the racialized other, and more about the overcoming of self-hatred built within communities of color as a response to the systemic and systematic oppression via whiteness (Yancy 2020, p. 165). This is a project taken up by various Black authors of Weird fiction, including in Victor LaValle's *The Ballad of Black Tom* (2016) and N.K. Jemisin's *The City We Became* (2020).

Before leaving "The Shadow Over Innsmouth," I want to point out that it is not hard to find explicitly racist language that alienates Asian communities as a dangerous Other.[9] In the opening pages, Lovecraft

[9] The story opens with a description of a government investigation in 1927–1928. The story takes place in between the Immigration Act of 1924 that excluded the immigration of people from Asian countries (The Immigration Act of 1924 (The Johnson-Reed Act) n.d.) and Order 9066, which incarcerated over 120,000 members of the Japanese and Japanese-American communities in the United States. While not explicitly a part of the story, the fact that this was a major legislation enacted just three years prior, establishes some racial context for the sense of alienness of the "Deep Ones."

describes Captain Obed Marsh as a "foreigner" with "South Sea islander" blood from his mother. We are told that an "epidemic" "brought from China or somewhere by shipping" ravaged the town in 1846 (Lovecraft, The Call of Cthulhu and Other Weird Stories 1999, p. 272). The inhabitants of Innsmouth are described as having "queer narrow heads with flat noses and bulgy, stary eyes that never seem to shut" (Ibid.). As in Bram Stoker's *Dracula*, the initial threat is a mixing of blood by a foreign source that would wreak havoc on the purity of the Western heritage (Stoker 1997). Upon my first reading of the story, I was brought face to face with my own sense of self-alienation. As a biracial, Japanese-American, I grew up with both a sense of self-hatred (and thereby a rejection of my "abnormal" half) and a feeling of alienation from the Asian-American community. By the end, the narrator is confronted with his own alien lineage and wavers between committing suicide and embracing both the "Deep Ones" and their underwater city. While the conclusion does not erase the egregious racism present at the beginning, it demonstrates the possibility of making racism a source of terror. Going against the fear of "outer, unknown forces" (Lovecraft 1973, p. 15), Lovecraft articulates a kind of wonder at embracing the other, which thereby upends the alien status of the racialized other. Whether or not this was the intention of the story, it demonstrates an escape from the trap of racism from within the structure of the Weird tale.

In their film analysis of *Pet Cemetery* and *The Ring*, Zefyr Lisowski argues that "oblivious in their own framing" of queer and disabled bodies from a supremacist perspective, horror stories offer a gift (Lisowski 2022, p. 49). Cast out of "the abled mainstream," they can create "a whole wellspring of culture and activism and vitality" (Ibid.). In some ways, this sense of the gift connects with the way that Yancy envisions his philosophical analysis of Whiteness as a gift. For Yancy, "books are not to be written for ourselves, though they often function as gifts that we would like to see make an important generative impact on shaping the thoughts, relationships, and lives of people who read them" (Yancy 2020, p. 3). In Lisowski's account, filmic images of monstrosity are a gift that enable an organization of community values, but require attention and thought to uncover. Yancy, offering a gift in response to the monstrosity thrust upon him via anti-Black racism, is working the opposite way. As an analytical tool, the eco-weird offers a kind of middle ground between the two. It uses the monstrosity and otherness established in Lovecraft's cosmic horror (of both race and land) to evaluate what is given and offer an opportunity to engage in the messy, and often monstrous, background of Lovecraft's sense of place.

To the second point raised at the beginning of this section, Lovecraft makes some comments that align him with a phenomenological account of place. While he does so from a specific perspective, there is a potential seed that would allow for a more robust understanding of how place is broadly generative for the Weird. In a 1930 letter to Derleth, he writes:

> I recognize the impossibility of any correlation of the individual and the universal without the immediate visible world as a background—or starting-place for a system of outward-extending points of reference. I cannot think of any individual as existing except as part of a pattern—and the pattern's most visible and tangible areas are of course the individual's immediate environment; the soil and culture-stream from which he springs, and the milieu of ideas, impressions, traditions, landscapes, and architecture through which he must necessarily peer in order to reach the 'outside.' (Lovecraft, Selected Letters III: 1929–1931 1971, pp. 220–221)

Sympathetic to a phenomenological reading, in this passage things emerge only against a background, and we cannot truly conceive of something that is not already part of some milieu or horizon. One question we must ask, of course, is who Lovecraft is describing when he says "individual"? Does he exclude the non-White community? If so, we could extend Lovecraft's reading of place and orient the familiarity of one's background as a foundation for the Weird in other places. That is, perhaps it is not specifically New England that is Weird, but New England as Weird for Lovecraft. It could also be the case that Lovecraft's New England was a specifically White version of New England. This seems most likely when recalling Menegaldo's distinction between the safe places of White Providence and the "morally inferior" non-White Providence. Likewise, we can imagine a familiarity with a different background, even within a similar geographical space, as a setting for other Weird horror tales. To this end, Yancy mentions or refers to "horror," "fear," and "terror" no less than 27 times in *Across Black Spaces*. Further, according to Yancy it is only through the "horror, messiness, and joy of human existence" that philosophy can take seriously the "importance of struggle against oppression through the exercise of human agency" (Yancy 2020, p. 141).

Also present in Lovecraft's articulation of the individual against a background is the intimate entanglement of environment and culture. The "soil" and the "culture-stream" work together to produce the background from which meaning emerges (Lovecraft 1971, pp. 220–221). Lovecraft points to these two components (soil and culture-stream) as the most visible aspects of a given context. However, this need not always be the case. By highlighting the "Old Providence," he identifies a specific culture-stream that he wishes to pull from. But some culture-streams are not easily identifiable and are, therefore, not the most visible parts of the milieu. This, perhaps, is where a phenomenological analysis differs from Lovecraft's analysis. Lovecraft assumes the visible portions of the background while phenomenological analysis works to make visible certain aspects of the background, throwing our intentional consciousness toward some narrowed plane of a broader horizon. In this sense, by reading Lovecraft's sense of place through a phenomenological lens we can make the background explicit as a foundational tool for analyzing the foreground. By so doing, the project of Weirding place takes on a different scope.

However, regardless of Lovecraft's explicit description of a visible background, the tension between the visible and the invisible is much more robust in his stories. We can take "The Colour out of Space" as a prime example. As previously stated, in the opening paragraph of this story, Lovecraft writes that "West of Arkham the hills rise wild." He goes on to establish a place, now crumbling, where "valleys and deep woods" have been untouched by human development and "the trees slope fantastically" (Lovecraft, The Call of Cthulhu and Other Weird Stories 1999, p. 170). What is remarkable about this opening passage is that not only does it establish some eerie natural landscape, but it pulls at the tension of what is given in perception and what is only alluded to. We can reframe this as a tension between presence and absence, which plays a major role in phenomenological analysis. We are offered a perspective of some untrammeled wilderness that has escaped the touch of any axe. We are absently present in a place that has eschewed the presence of human involvement. Further, we are told that the place has turned "foreigners" away because the "place is not good for the imagination, and does not bring restful dreams at night" (Ibid.). But this turning away is not because New England is specifically a White space (it may be, but that is not what is necessarily at play here). Rather, the trees themselves do not seem to fit, with trunks "too big for any healthy New England wood" (Lovecraft 1999, p. 171). Here, the environment itself is made Weird, both inviting us in and pushing us out.

This environmental Weirding continues after the narrator hears Ammi Pierce's story. As he escapes back to his hotel, we are told of the "dim chaos of old forest and slope" and the "grey blasted heath where the black well yawned deep beside the tumbled bricks and stones" (Lovecraft, The Call of Cthulhu and Other Weird Stories 1999, p. 173). But more than the forest itself, it is the water that will cover the area that conveys the scope of the environmental terror. As the new reservoir is set to bury this uncanny place with its "sinister starts," the narrator claims that he could not be paid "to drink the new city water of Arkham" (Ibid.).

Lovecraft goes on to generate a sense of the Weird that is closely associated with the land. First, the plant crops begin to fester with an abundant color that is only a color by analogy. Then the livestock grows too large, their meat no longer edible. Here the use and value of the land is disrupted, spoiled, and the foundation for the very reason for having a farm at all is upended. Last, the humans begin to go mad. First, the women, then the boys, and finally the men. It is as if Lovecraft is working his way up a patriarchal "great chain of being." But because he is a misogynist and a strict materialist, there is nothing higher than men in his chain. What is fascinating about this chain of events is that Lovecraft is Weirding the foundations of capital that were largely driven by his heritage of Whiteness.

Brian Williams and Jayson Maurice Porter identify cotton as an agricultural example of Whiteness at work on the land and on the bodies of people of color. According to them, the industrialization of cotton not only worked to destroy the land via the oppression, but it was deeply tied to the use of slaves as property. This was not a historical problem, but the roots of the capitalist context in which we continue to work. For Williams and Porter, the roots of capitalism in cotton production "shape(s) the political ecologies of contemporary capitalism: racialized labor systems, the uneven valuation of bodies, and the theft and degradation of land are all central to the realization of value and accumulation of capitalist value, and the violent displacement of crises of pollution and degradation" (Williams and Porter 2022, p. 502). Putting this into the context of "The Colour out of Space," we can recall the ongoing poisoning of the land and the effect this has on the family. Williams and Porter highlight how Black slaves in the U.S. South were not only required to participate in the destruction of the land via cotton harvesting but also employed to collect guano. The collection of guano meant that Black slaves were brought intimately close to "clouds of guano dust full of ammonia and nitrates" that "enveloped indentured workers, leading to nosebleeds, blindness, and lung damage" (Williams and Porter 2022, p. 507). By bringing the danger of the oppressed land to the White farm family, Lovecraft unintentionally leaves room to read "The Colour out of Space" as a mechanism of departure, whereby the assumed hegemonic normalcy of Whiteness is challenged.

If his sense of place is tied to his racist and xenophobic beliefs, then his sense of place is also tied to the kind of environmental destruction that is at play in "The Colour out of Space." Whiteness produces environmental destruction through the absolute use of the land, including some parts of the human community considered as extensions. We can recall Aldo Leopold's introduction to the land ethic, where he describes Odysseus' slave girls as an extension of the land. Their use and destruction rendered acceptable because they were mere property (Leopold 1968, pp. 201–202). It is not hard to tie this to the practice of slavery that built wealth early in the United States. We can also think of the destruction of the land by factory farming practices that produce excessive waste and pollute local watersheds. Not only do we have a Weirding of land through its destruction via "development," but we have a destruction of the water source, which is a lasting impact of the alien intrusion in "The Colour out of Space." As the story ends, we are told that the new reservoir will cover over the "blasted heath" and the land that used to be occupied by the Gardner family. For the narrator, this indicates that the purpose of the reservoir is self-defeating as the water will become tainted by the alien force that destroyed the land's usefulness. Here, we have the culmination of the eco-weird in its ecological rather than its "cosmic" scale. The threat does not come from the outside as an intrusion but is an extrusion of what is already in place. The acts of dominating capitalism, land use, and Whiteness all show up as the source of the Weird that deterriotoralizes and thus leads to our ultimate departure. The question, and it may be a normative one, is whether this deterritorialization is a departure from the Weirdness of our own context.

Weirding the familiar, which thus calls into question the familiar, offers an opportunity to explicitly highlight our own background in hopes that it moves us along to an inevitable departure. But Weirding the familiar does not necessarily lead to any specific place. If there is an ethical goal attached to the Weird, it would be highlighting how deterritorializing the familiar enables us to move beyond the given to the ungiven, beyond the possible to the impossible. While the eco-weird is not specifically prescriptive, it makes it possible to generate a value set that at least foregrounds the environment, the destructive background of hegemonic Whiteness, and enables some other option. In my eco-weird analysis of Lovecraft's sense of place, I have read Lovecraft's writing through phenomenology, which has specifically highlighted the way that place, race, and environmental values are deeply entangled and thus require a set of tools that allow us to think through them together. By providing the groundwork, it becomes an open task to generate a better version of our sense of place that avoids the hegemonic racism implicit in Lovecraft's stories and explicit in his writing.

Acknowledgments I would like to thank the John Hay Library at Brown University for awarding me the S.T. Joshi Endowed Research Fellowship, which made this chapter possible.

References

Ashley, Mike. 1988. The Cosmic Connection. *Crypt of Cthulhu* 7 (7): 3–9.

Brown, Charles S., and Ted Toadvine. 2003. Eco-Phenomenology: An Introduction. In *Eco-Phenomenology: Back to the Earth Itself*, ed. Charles S. Brown and Ted Toadvine, ix–xxi. Albany, NY: SUNY Press.

Cisco, Michael. 2021. *Weird Fiction: A Genre Study*. Cham: Palgrave Macmillan.

Harman, Graham. 2012. *Weird Realism: Lovecraft and Philosophy*. Washington: Zero Books.

Jemisin, N.K. 2020. *The City We Became*. New York: Orbit.

Joshi, S.T. 1990. *The Weird Tale*. Austin, TX: University of Texas Press.

———. 2008. *The Rise and Fall of the Cthulhu Mythos*. Poplar Bluff, MO: Mythos Books.

———. 2013. *I Am Providence—The Life and Times of H.P. Lovecraft: Volume 2*. New York: Hippocampus Press.

Klein, T.E.D. 1965. A Dreamer's Tale. In *Dagon and Other Macabre Tales*, xiii–lii. Sauk City, Wisconsin: Arkham House Publisher's Inc.

LaValle, Victor. 2016. *The Ballad of Black Tom*. New York: Tordotcom.

Leopold, Aldo. 1968. *A Sand County Almanac: And Sketches Here and There*. New York: Oxford University Press.

Lisowski, Zefyr. 2022. The Girl, The Well, The Ring. In *It Came from the Closet: Queer Reflections on Horror*, 41–50. New York: Feminist Press.

Lovecraft, H.P. 1965. Autobiography: Some Notes on a Nonentity. In *Mirage on Lovecraft: A Literary Review*, ed. Jack L. Chalker, 3–10. Baltimore, Maryland: Jack L. Chalker and Mark Owings: Publishers.

———. 1971. *Selected Letters III: 1929–1931*. Edited by August Derleth and Donald Wandrei. Sauk City, Wisconsin: Arkham House: Publishers.

———. 1973. *Supernatural Horror in Literature*. New York: Dover.

———. 1982. Notes on Writing Weird Fiction. In *H.P. Lovecraft: Uncollected Prose and Poetry 3*, ed. S.T. Joshi and Marc A. Michaud, 42–44. West Warwick, Rhode Island: Necronomicon Press.

———. 1999. *The Call of Cthulhu and Other Weird Stories*. Edited by S.T. Joshi. New York: Penguin Classics.

Mariconda, Steven J. 1986. Lovecraft's Concept of "Background". *Lovecraft Studies* 12: 3–12.

———. 1987. On the Emergence of "Cthulhu". *Lovecraft Studies* 6 (2): 54–58.

———. 1991. Lovecraft's Cosmic Imagery. In *An Epicure in the Terrible: A Centennial Anthology of Essays in Honor of H.P. Lovecraft*, ed. David E. Schultz and S.T. Joshi, 188–198. Madison, NJ: Fairleigh Dickonson University Press.

Meillassoux, Quentin. 2008. *After Finitude: An Essay on the Necessity of Contingency*. New York, NY: Continuum.

Menegaldo, Gilles. 1981. The City in H.P. Lovecraft's Work. *Lovecraft Studies* 1 (4): 10–19.

Merleau-Ponty, Maurice. 1962. *The Phenomenology of Perception*. Translated by Colin Smith. New York, NY: Routledge Classics.

Onishi, Brian. 2020. Weird Ecologies and the Uncanny in The Happening. In *Philosophy, Film, and the Dark Side of Interdependence*, ed. Jonathan Beever, 157–173. Lanham, MD: Lexington Books.

Onishi, Brian Hisao. 2023. *Weird Wonder in Merleau-Ponty, Object-Oriented Ontology, and New Materialism*. Cham: Palgrave Macmillan.

Schultz, David E. 1991. From Microsm to Macrocosm: The Growht of Lovecraft's Cosmic Vision. In *An Epicure in the Terrible: A Centennial Anthology of Essays in Honor of H.P. Lovecraft*, ed. David E. Schultz and S.T. Joshi, 199–219. Madison, NJ: Fairleigh Dickinson University Press.

Stoker, Bram. 1997. *Dracula*. Knoxeville: Wordworth Editions.

The Immigration Act of 1924 (The Johnson-Reed Act). n.d.. https://history.state.gov/milestones/1921-1936/immigration-act. Accessed August 19, 2023.

Trigg, Dylan. 2012. *The Memory of Place: A Phenomenology of the Uncanny*. Athens, OH: Ohio University Press.

Williams, Brian, and Jayson Maurice Porter. 2022. Cotton, Whiteness, and Other Poisons. *Environmental Humanities*: 499–521.

Yancy, George. 2020. *Across Black Spaces*. Lanham, MD: Rowman & Littlefield.

CHAPTER 4

The Weird as Crisis Genre: Tipping Points, Ontological Reorientation, and the Desert Tide in Algernon Blackwood's "Sand" (1912)

Emily Alder

4.1 Introduction

Something planetary is breaking through, entailing a politicization of what may once have seemed insignificant.

—Timothy Clark (*Ecocriticism on the Edge: The Anthropocene as Threshold Concept*)

All that was free, untied, responded instantly like chaff; loose objects fled towards it; there was a yielding in the hills and precipices; and even in the mass of Desert which provided their foundation.

—Algernon Blackwood ("Sand")

E. Alder (✉)
Edinburgh Napier University, Edinburgh, Scotland
e-mail: Em.Alder@napier.ac.uk

B. H. Onishi, N. M. Bell (eds.), *The Call of the Eco-Weird in Fiction, Films, and Games*,
https://doi.org/10.1007/978-3-031-77126-2_4

The eco-weird recognises agencies that extend through and beyond the matter of the world known by humans. In the subject of this chapter, Algernon Blackwood's short story "Sand" from the collection *Pan's Garden* (1912), protagonist Felix Henriot glimpses a fleeting manifestation of ancient forces poised to materialise in the human realm through the medium of sand. This story exemplifies Blackwood's recurrent interest in encounters between human and inhuman realms and in the immanence of numinosity. It does so through a characteristic feature of weird tales: hearing the "scratching of outside shapes and entities on the utmost rim of the universe", as H. P. Lovecraft (1973, n.p.) put it, facing the moment of breaching the boundaries of what was thought to be the known world.

When weird tales demand ontological reorientation like this, they reveal inherently ecological qualities. In this case, "Sand" contrasts the vitality of the desert to the conditions of British modernity, particularly its materialism, colonialism, and anthropocentricism. The experiences undergone by Felix Henriot suggest human and other-than-human realms overlapping and co-existing, holding out the possibility of greater ecological understanding even in the face of terrifying forces, while around him human greed, arrogance, and indifference work to close off such potentiality. At its crux, the narrative reaches a tipping point at which conventional understanding of human reality is almost swept away, as powers provoked by an occult ceremony in the night-time desert threaten a rapid and overwhelming ascendancy.

Felix identifies the desert power as "the Ka of Ancient Egypt" (Blackwood 1912b, p. 271); however, rather than approaching the story through Egyptology, my eco-weird reading of "Sand" looks to the matter of stone and sand and to the climate transformations of the Eastern Sahara in deep time. In Blackwood's story, sand links the genre conventions of the Weird—the encounter with the numinous beyond the everyday—to the Anthropocene. Sand indexes the slow passage of geological time as well as the swift unpredictability of climate change effects, endowing the desert's other-than-human existence with agency of its own. Through Felix, the narrative models an ontological reorientation to an eco-weird reality of inhuman temporal and spatial scales and agencies, a reorientation based on the knowledge of intuition or spirit as much as that of science. Through characters like Richard Vance, however, "Sand" is also cynical about impacts of human actions; aware of climate's capacity for change, the story also understands that human interference in the biosphere can have drastic consequences.

The Weird, then, has much to offer ecocritical reading. As Timothy Clark's remark in my epigraph suggests, today's response to entering the Anthropocene—which describes the ongoing period in which human activity has become the dominant force affecting Earth systems[1]—may require a "planetary" shift, a vast-scale reorientation of how we understand ourselves as part of a greater whole. Earlier in the same passage, Clark (2015) suggests that the concept of the Anthropocene

> puts in crisis the lines between culture and nature, fact and value, and between the human and the geological or meteorological. As a bewildering and often destructive contamination of human aims and natural causality, the Anthropocene manifests itself in innumerable possible hairline cracks in the familiar life-world. (p. 9)

Clark's remarks point to the ontological reorientation required by the Anthropocene as a concept. Weird tales, adept at putting established boundaries, scales, and causalities under pressure, yield very well to Anthropocene reading. The Anthropocene may be a modern term, but many features of the continuing ecological catastrophe are tied to the development of industrialisation, globalised capital, and Enlightenment thought since especially the eighteenth century. While modern data demonstrating the anthropogenic causes of global warming were not available at the time Blackwood was writing, the concept of climate change was long known, as was the capacity of human activity to degrade natural environments. From our twenty-first century perspective, we recognise that the causes of ecological and sustainability crises making their presence felt now were well underway by the end of the nineteenth century. In this context, weird tales reward consideration, including Blackwood's, whose fiction represents one of the best test cases I know for the eco-weird in the early twentieth century.

4.2 Blackwood and the Eco-Weird

The Weird's capacity for ecological reorientation derives partly from its ontological position in relation to otherness that differs from the Gothic's familiar returns. For China Miéville (2012), the Weird "works through radical otherness, a counterposing alterity" and is "an enmonstered iteration of a long, strong aesthetic and philosophical tradition, one endlessly obsessed with questions of the Awesome, a beauty that is terrible and

[1] For a full discussion of the Anthropocene, see, for example, Lewis and Maslin (2018); and in relation to literature, see Menely and Taylor (2017).

beyond-kenn-or-kennableness" (p. 380). Ontological reorientation endows the Weird with radical, transformative potential. In context of political 2000s "New Weird" fictions, for example, Miéville suggested that "the neo-liberal claim that There is No Alternative is all about trying to draw the line of the 'never-possible' at a place which strips humans of any meaningful transformative agency" (Gordon and Miéville 2003, p. 368). The Weird, however, can narrativise transformative possibilities for human characters facing what seem like the relentless movements of modernity, possibilities that are always in tension with the limits to individual agency that threaten to, and sometimes do, overwhelm them.

Here, I argue the same principle applies to the eco-weird. Firstly, the climatic and ecological catastrophe we are living in is closely tied to the progress of liberalism and neoliberalism. Secondly, response to ecocatastrophe can be paralysis as individuals struggle to comprehend meaningful action in relation to the economic, political, and social engine of the Capitalocene (Moore 2015). It is a challenge to exert individual agency within the giant systems that drive, and could mitigate, anthropogenic climate change, as it is to exert oneself against the vast alterities that often figure in weird tales. In weird tales, alterity beyond human knowability may be overwhelming or undefeatable, but it is also ecologically powerful, rupturing illusions of anthropocentricity, revealing unthought-of possibilities, and leading to transformation. Weird fictions position subjectivity precariously, daring us to think differently, understand differently, act differently, even, or especially, if it takes us somewhere uncertain, threatening, and terrifying. This kind of response, perhaps, is what climate crisis demands.

Blackwood's fictions yield to an eco-weird analysis when they make such demands, as they often do, requiring characters, and readers, to reposition themselves in relation to the rest of world and to how they know it. Blackwood himself had a more capacious understanding of the nature of the universe than was proposed in scientific philosophy, religion, or occultism. Nineteenth-century positivist science, rooted in Enlightenment philosophy, more or less depended on an anthropocentric assumption of human mastery over nature—human reason and the scientific method had the capacity to reveal and understand the secrets of the universe.[2] For Blackwood this was too limiting. Secrets and wonders remained in the

[2] See my *Weird Fiction and Science at the Fin de Siècle* (Alder 2020) for examination of the Weird's relationship with changes in scientific thought and knowledge in the period.

world and were, and should be, beyond human understanding. Yet to accept this truth means abandoning ideas of human supremacy, control, and centrality in the world that are based on dualistic conceptions of human–nature relations. In "Sand", for example, there is little if any distinction in Felix's perception between desert existence and human existence. In Blackwood's eco-weird, Cartesian boundaries such as those between scientist/subject, intellect/body, culture/nature, or material/spiritual are not so much collapsed as non-existent. The more-than-human world extends through and beyond the quotidian, visible physical realm defined by empiricism and materialism; Blackwood did not believe in the supernatural so much as know that all phenomena, however wondrous, were natural (Ashley 2001). On this basis, many of his stories are positioned ready to construct or at least reach towards a non-anthropocentric set of ecological ethics.

Blackwood's stories frequently depict characters overwhelmed and transformed by their encounters with ecological alterity, including in some of his best-known tales such as "The Wendigo" (1910) and "The Man Whom The Trees Loved" (1912a). David Punter (2013) and Michelle Poland (2017) have both argued that Blackwood's tales overturn anthropocentrism by diminishing the importance of the human, recognising other forms of agency, being, and intrinsic valuing of non-human and non-animal entities. For John M. Miller (2020), repositionings like these are produced by the Weird's formal technique of dwelling on descriptions, especially of landscapes; in his example, Blackwood's "The Willows" (1907), "those parts of the narrative that should provide a static background to human events prove riddled with one or more disruptive entities", prompting ecological realisations (p. 247). In Blackwood's "becoming-forest" stories, including "The Man Whom The Trees Loved", Punter sees the natural world "continually spilling over and spilling out, proliferating in ways which Deleuze and Guattari would refer to as 'rhizomatic', overwhelming attempts at order" (pp. 47–48). In Blackwood's 1911 novel *The Centaur*, this process is almost literally the case: rhododendrons veil and reveal the earth consciousness and ***urmensche*** to Terence O'Malley. Rhizomatic proliferation subverts the visible, predictable hierarchy of the tree of life; rhizomatic energies infiltrate the world under the surface and emerge everywhere, just as Blackwood's infusive, effusive vision of human–nature union expresses it.

Poland concentrates on the collection *Pan's Garden*, in which "Sand" appeared. For Poland, "Blackwood's work effortlessly builds a discourse around a shifting natural world" (p. 57). Drawing on Val Plumwood, Poland argues that "the really subversive element in Pan's garden is one that opens the gate of 'human self-enclosure' and enters into a radical immersion with the more-than-human world" (p. 61).[3] The collapse of physical and spiritual boundaries in these stories' environmental spaces produces a new understanding of the cosmos and the human place in it. Thus, in "The Man Whom The Trees Loved" and "The Heath Fire" (2012), Bittacy and O'Hara become one with trees and flames, respectively. However, unity is not reached through peaceful harmony with a passive, static natural world, but instead through recognition of nature's own agency, otherness, and unpredictability as something "turbulent and mysteriously erratic" (Poland 2017, p. 66). By recognising the autonomous, changeable, and damaged qualities of the natural world, Blackwood's stories speak well to our contemporary moment of environmental crisis, in which, for example, effects of global warming will not be predictable, even-handed, or necessarily mitigable.

Important in these stories is not only the process of immersion but also what one is being immersed *in*: not in a benign harmony with "nature" but in a difficult though not necessarily undesirable relationship with a demanding cosmos. In *The Centaur*, O'Malley profoundly connects with a living more-than-human, more-than-visible world. His expanding awareness of "the collective consciousness of the Earth" (p. 199) is at times couched in terms of energy flow:

> The cosmic life ran through his being, lighting signals, offering service, more—claiming leadership. With it, however, came no loss of individuality, but rather a powerful increase of life by means of which for the first time he dreamed of a fuller existence. (p. 136)

Immersion in the life of the cosmos, in the consciousness of the earth, is potentially overwhelming, but identity is not lost; rather, it grows more complete, reaching a "fuller existence". O'Malley's transformation is not a transfer from physical to immaterial realm but a reinvigoration on a bodily and material as well as spiritual level; he "experience[s] an immense, driving push upon what Bergson has called the *élan vital* of his being"

[3] Poland quotes from Val Plumwood, *Environmental Culture: The Ecological Crisis of Reason* (London: Routledge, 2002).

(p. 136).[4] Through Henri Bergson's *élan vital*, a physical force explaining the life of matter, *The Centaur* insists on the validity of both spiritual and empirical knowing of the world and of the numinous as immanent.

In these ways, Blackwood's eco-weird stories model ontological reorientation based on epistemologies that validate additional kinds of knowledge to those accepted by mainstream science, such as the spiritual and extra-sensory. Weird fiction's ability to make ecological responses like this is partly due to the conditions of knowledge-making at the time of its emergence in the late nineteenth century. Weird fiction cohered into a distinct form at the same time as new scientific epistemologies were developing as alternatives to the positivism dominating most of the nineteenth century (Alder 2020). For example, much of the *fin-de-siècle* occult revival, in its many forms, depended on a process of naturalising the supernatural, accounting for an unseen world through epistemologies that challenged or diverged from conventional nineteenth-century positivism, even if they didn't intentionally set out to do so. In 1903, psychical researcher Frederic Myers wrote that "the time is ripe for a study of unseen things as strenuous and sincere as that which Science has made familiar for the problems of earth" (p. 275). The problem, however, was that "the kind of universally accepted evidence that he and other members of the Society [for Psychical Research] felt it essential to discover was precisely what eluded them" (Beer 1998, p. 186). Weird tales often attempt to reconcile this epistemological contradiction: empirical proof was required, but positivism was inadequate for knowing what was an essentially anti-positivist universe.

Weird tales offer scenarios that are open to radical kinds of knowledge based on alternative forms of evidence. Blackwood's "The Willows", I have argued elsewhere, engages established and emergent theories and discourses of physics and occultism to demonstrate the nature of a weird reality in which both positivist and anti-positivist explanations or framings may conflict yet co-exist as valid worldviews (Alder 2020). These questions become ecological because the knowledge and belief positions contested by the Weird are tied to the damaging consequences of Cartesian binary thinking that has led to the ongoing ecological crisis. Western thought is still entrenched in hierarchical ideas of human–nature separation and of things of nature as resources available for human use. These ways of thinking underpin unsustainable practices like continual economic

[4] See also Ashley (2003, p. 3).

growth and energy consumption which lead to environmental degradation and global warming—practices that are emotional and social as well as political, economical, and technological. Achieving sustainability requires a cross-disciplinary response in which traditional scientific positivist methods are not the only ones required. In this sense, sustainability thinking has much in common with the eco-weird, which can model cross-disciplinary responses in narrative form. Weird tales in the period in which Blackwood was writing took issue with the predominate worldviews and methodologies of the *fin de siècle,* and may have much to contribute to ways of thinking about today's climate crises and responses to them.

4.3 The Weird as Crisis Discipline

In the post-war period, along with realising the anthropogenic causes of environmental catastrophes came the realisation of their multidisciplinary implications (Egan 2017). In the 1980s, Michael Soulé proposed the necessity of developing "crisis disciplines" as the most appropriate response in science. The term "describes the changing nature of scientific praxis in the face of ecological disaster" (Egan 2017, p. 27). Crisis disciplines are interdisciplines such as conservation biology or environmental communication that work to respond urgently to issues like biodiversity loss by recognising the mutual relationships between, for example, society and ecosystem stability (Meffe 2001); conservation science "acknowledge[s] the importance of culture, social context, and human psychology in forming effective policy and embrace the humanities and social and biological science" (Stirling and Burgman 2021). For Soulé (1985), conservation biology and other crisis disciplines were "synthetic" and "multidisciplinary", and "holistic" in the sense that "multidisciplinary approaches will ultimately be the most fruitful" (p. 728).

Crisis disciplines as Soulé conceived them depended on epistemologies that could bring different disciplines together and productively reconcile their conflicts and divergences: "In crisis disciplines", he wrote, "one must act before knowing all the facts; crisis disciplines are thus a mixture of science and art, and their pursuit requires intuition as well as information" (p. 727). To address a "looming threat", a crisis discipline "must offer recommendations for management or intervention to protect imperiled species, biological communities, or ecosystems, under conditions of urgency and often without theoretical or empirical guarantees" (Cox

2007, p. 6). The Weird, I suggest, resembles a crisis discipline in that it blurs traditional boundaries, participates in multiple genres, accepts uncertainty, and acts on intuition to address unanswerable questions. In Blackwood's work, the supreme skill of "physician extraordinary" John Silence in "A Psychical Invasion" (1910), for example, rests on both his intuition and his training. As a prospective client puts it, his "sympathetic heart" and his "trained psychic knowledge" (p. 1) are what enable him respond effectively to crisis after crisis and "bring balance to a world that threatens to fall out of sync" (De Cicco 2012, p. 19). Weird tales also narrativise recommendations, of sorts, for regaining stability; it may be a temporary, conditional, fragile, compromised, or even elusive stability, but nonetheless they often model an alternative to crisis, or suggest a route through it.

These are narratives needed in the Anthropocene too. For Egan (2017, p. 27), the new forms of scientific enquiry that arose in crisis disciplines from the 1970s constituted "a new science of survival". Crisis disciplines are fields in which "practitioners typically seek solutions before a situation becomes irreversible" (Stirling and Burgman 2021, p. 1739). The eco-weird can also be seen in this way, in tales like "Sand" that present a crisis threatening the known way of being in the world and which elicit an urgent response that depends on a rapid revision of one's worldview and basis for knowledge. To put it another way, the response of crisis disciplines to anthropogenic ecological catastrophe resembled the way Blackwood's fiction responded to the crisis he detected in how modernity thought it knew and existed in the world. The capacity to detect and engage anthropogenic environmental problems, even before they were recognised as such, exists in the Weird through its function as a "crisis genre", a crisis discipline in narrative form.

"Sand", as I will explore, suggests ways of overcoming dualisms and validating new forms of knowledge, while remaining aware of the potential effects of interference with the biosphere. The story's distinctively phenomenological narration becomes a route to gaining fuller knowledge of the other-than-human, in this case the desert. Phenomenology enables the story to engage with the possibility of climatic change over vast geological timescales as well as alarmingly short ones, by validating the life of the desert and the numinous it conceals as real. It suggests the scale of the ontological reorientation that may be necessary to reach a new provisional stability after anthropocentricity has fallen away, as well as the scale of the challenges of getting there. As a story belonging to the crisis genre that is the eco-weird, "Sand" demonstrates how spiritual and phenomenological knowing along with geological knowledge can yield meaningful ecological insights.

4.4 The Desert Tide: Blackwood's "Sand"

"Sand", like another Blackwood story, "A Descent into Egypt" (1914), was inspired by a visit the author made to Egypt in 1912. It is the story of Felix Henriot, a young man who seeks escape from British urban dullness. He is fascinated by what he perceives as the hidden power of the desert, in contrast to the shallow knowledge of the popular Egyptology of the time. Spending time in the town of Helouan (Helwan) on the edge of the Eastern Sahara, Felix makes the acquaintance of Richard Vance and Lady Statham. Lady Statham appears to possess the secret knowledge Felix yearns for, which draws him into participation in an occult ceremony in the nearby Wadi Hof whereby she attempts to secure a path for pandimensional powers into the human world. Her experiment nearly succeeds, but she is betrayed by Vance at the point of breakthrough.

Felix, "caught" by the "syren's call" of adventure, travels to Egypt to escape the trappings of modern London, its "shops and crowds and motor-'busses'" (p. 227). The opening pages establish a sharp binary between the dreary, fog-stifled city and Felix's romantic daydreams of the Mediterranean and Aegean: "into his stuffy room, slipped the singing perfume of a wall-flower on a ruined tower [...] The darkness melted from the city. The fog whisked off into an azure sky. The roar of traffic turned into booming of the sea" (p. 227). Felix's fantasies of these realms of adventure establish a familiar contrast between the imperial metropole and its colonised or occupied lands (Egypt was under British control at the time of the story's writing and setting) that pose the latter as spaces available for colonial use. These lands are othered and exoticized in Felix's daydreams, constructed in terms of their landscapes, climate, and mythology, in contrast to the social, quotidian details of London life.

Such dualistic othering has ecological implications, resulting in, as Val Plumwood (2003) puts it, "a polarized understanding in which overlap and continuity between the human and non-human spheres are denied and discouraged" (p. 59). Plumwood sees a close connection between dualistic attitudes towards people and more-than-human others, and those towards place, attitudes which owe much to British colonialism and global commodity culture. These divisions are particularly problematic in context of a globalised economy in which the developed lifestyles of Western or Global North societies rely on "disregarding the many unrecognised, shadow places that provide our material and ecological support" (Plumwood 2008, n.p.). In "Sand", the desert itself acts as a shadow place; the lifestyle of the tourists and colonial British residents in Helouan depends on it, but its vitality is invisible to them. Blackwood, like fellow Weird writer Arthur Machen, saw problems with modernity's predominantly materialist worldview that failed to recognise the greater life of the cosmos. Those problems owe much to anthropocentricism and the spread of consumer capitalism, and so are related to the causes of climate change and environmental degradation.

In "Sand", Felix himself undergoes ecological growth; he gradually attains a distinct understanding of the desert as alive, perceiving its continuity and overlap with himself. But he also recognises its incompatibility with his social world, and the desert in Felix's perception responds antagonistically to the shallowness of a materialist, colonialist, anthropocentric modernity that proceeds with little regard for a wider more-than-human world. The story also poses a scenario of human interference with the land through Lady Statham's ceremony of evocation, intended to stimulate the desert's shapeless powers to take material form through sand. The story's representations of rocks and sand evoke the geological timescales on which their evolution and transformation normally depends, accelerated to a tipping point whereby they threaten to overwhelm the human world.

The concept of the "tipping point" dates back to the 1880s, and the term later spread through multiple disciplines including social and natural sciences (Milkoreit et al. 2018). In earth science, a tipping point can be defined as "any situation where accelerating change caused by positive feedback drives the system to a new state" (van Nes et al. 2016, p. 904). In the context of climate change, a tipping point is a potential result of breaching a critical planetary boundary such as the safe limit of greenhouse gases in the atmosphere (Rockström et al. 2009). One of the threats at the heart of "Sand" is the potential for breaching safe limits because of human interference in the biosphere. If for Blackwood the natural and supernatural amounted to the same thing, then Lady Statham's experiment extends human interference in the known biosphere into the numinous that lies beyond it, with no guarantee of controlling the outcome. While the term "tipping point" has other disciplinary meanings and applications, the sense in which I use it here is primarily ecological, describing the "phenomenon of regime shifts in ecological systems" (Milkoreit et al. 2018, pp. 1–2). A contiguous spiritual and ecological regime shift is essentially what Lady Statham is attempting to generate, and the story also offers a glimpse of what its consequences could be.

In drylands, one such ecological regime shift is desertification, and the way Felix observes and experiences the vitality of the desert has much in common with processes of desertification. Desertification has numerous definitions, and the types, mechanisms, and effects of desertification vary from region to region.[5] However, typically it is characterised by changes in plant cover, soil erosion, salinisation, and other indicators. Desertification is not inherently anthropogenic in cause. The Eastern Sahara's most recent period of aridification began 5000–7000 years ago following climate oscillation and sea-level rise in the Pleistocene, but it had undergone many evolutions in the millions of years before that (Zaki et al. 2020). However, in recent centuries a combination of methods of land management along with effects of anthropogenic climate change have influenced its increase. Desertification and droughts are indicators of global climate crisis, and are linked to other global environmental disruptions and tipping points; for example, disruption to the AMOC (Atlantic meridional overturning circulation) caused by melting ice sheets, which is itself approaching a tipping point, will increase Saharan desertification (Lenton et al. 2019; Ditlevsen and Ditlevsen 2023). In the context of drylands, "more than climatic contingencies, human interference in the biophysical environment causes degradation and desertification" (Rodrigues do Nascimento et al. 2023, p. vi). Finally, desertification is not only caused by global climatic changes; it also has a global impact in return as "[t]he long-range effects of desertification modify climate, global biochemical cycles, and human geography" (D'Odorico et al. 2013, p. 337). The effects of desertification environmentally also have consequences agriculturally and socially.

Desertification is recognised to be not a fixed state but a "dynamic process that includes both biophysical and socioeconomic factors", with the boundaries of deserts being variable (Verón et al. 2006, p. 754). Other studies suggest that desertification may be a process of transition between two states, meaning that "if disturbed beyond a critical threshold"—a kind of tipping point—"these systems move towards the alternative stable state of degraded land" from which it is difficult to return (D'Odorico et al.

[5] For detailed overviews of desertification, see, for example, Rodrigues (2023) and D'Odorico et al. (2013).

2013, p. 229). Despite today's recognition of desertification as dynamic and dependent on many factors not all within human control, well into the mid-twentieth century it was predominantly understood as a state caused by human activity (it therefore followed that human actions, for example around planting and farming, could and should redress it). It was understood as an expansion of desert boundaries, "spread[ing] from desert cores by means of 'sand invasions' and, in general, these changes were considered irreversible" (Verón et al. 2006, p. 755). This earlier, discarded perspective on desertification is important for "Sand"; Felix's experiences of the desert's life often appear as "sand invasions" from deep in the desert, into the city and even into his hotel.

If responding to desertification and its implications requires the interdisciplinary approach of a crisis discipline, "Sand" fictionalises a way of comprehending them through spiritual or intuitive as well as empirical knowing. The narration through Felix's perspective is highly phenomenological, defined as a method of gaining knowledge about the world through the way it appears in conscious experience (Gallagher 2022). In the story, events and interactions are always channelled through Felix's experience of them. What he experiences is a desert expanding its boundaries, an incursion into the human world that is both terrible and exhilarating. Early on in his stay in Helouan, Felix wakes "with an appalling feeling that the Desert had come creeping into his room and now stared down upon him where he lay in bed" (p. 244). On this occasion, "[i]t seemed next day that he had dreamed it. Something enormous, with rustling skirts of sand, had just retreated far into the Desert. Sand went with it—flowing, trailing, smothering the world" (p. 244). The Desert revisits him persistently, however, and he ceases to question its reality, which comes to be more real to him than the world of the hotel. Through his new knowing of the desert, Felix is better positioned to face the potential implications of its spread and its agency.

On another evening, while speaking to Lady Statham, who evokes associations in him of the desert and perhaps of a previous life, Felix "felt the Desert had come close. Into this tent-like hall of the hotel came the sifting of tiny sand. It heaped softly about the very furniture against his feet, blocking the exits of door and window" (p. 273). Characteristically, to experience the desert's matter is also to experience its time; the sand not only blankets the interior of the building but also "shrouded the little present. The wind that brought it stirred a veil that had hung for ages motionless..." (p. 273). On this occasion, Felix's social and temporal reality ceases operation entirely, becoming diminished and "little", "shrouded" and ghost-like, while sand, he comes to realise, will form the bodily matter of the desert powers.

That these experiences are not personal hallucinations is established by the experience of the niece of Felix's friend John Mansfield, which John reports in more-or-less empirical terms:

> The room was full of—of things, yet she could see nothing. She *felt* them, you see. [...] she said it made the door bulge inwards from the next room, but not the door alone; the walls bulged or swayed as if a huge thing pressed against them from the other side. [...] She said it was all *one* thing—trying to get in; just as water, you see, would rush in through every hole and opening it could find, and all at once. And in spite of her terror—that's the odd part of it—she says she felt a kind of splendour in her—a sort of elation. (p. 287)

Mansfield's struggle to articulate what happened is marked by a number of awkward phrases, hesitations, and dashes that interfere with coherent description. This passage contrasts with the flowing, sensuous language used for the Desert's visit to Felix in the hotel, which positions Felix as an integral part of the experience rather than its observer. Mansfield's clumsy account—observations of Lady Statham's early experiments in the hotel room next to that of John's niece—supports Felix's experiences, but is actually harder to make sense of than his phenomenological perceptions.

While Felix tunes in with the desert, suggesting that other, more open ways of relating to the other-than-human are possible, he is surrounded by a modernity focused on the material and the empirical. Lady Statham claims that "[t]he mind to-day wears blinkers, studies only the details seen directly before it" (p. 266). Felix's phenomenological experiencing of the desert is presented as more successful and complete, leading to valid knowledge. In phenomenology, "the person who opens her eyes and sees the world, sees it and has knowledge of it consciously" (Gallagher 2022, p. 2). Consciousness is how human beings access reality, and accordingly when Felix asks Lady Statham to what "useful end" her experiments tend, she replies, "[t]he extension of spiritual knowledge and the widening of life" (p. 277): the expansion of consciousness is a route to increased knowledge.

By contrast, most of the modern world has neglected the true spiritual (and therefore also natural) realm of knowledge. The other people in Felix's hotel in Helouan are unable to recognise the living power of the desert, and he wonders

> how people dared their little social activities within its very sight and hearing; how they played golf and tennis upon reclaimed edges of its face, picnicked so blithely hard upon its frontiers, and danced at night while this stern, unfathomable Thing lay breathing just beyond the trumpery walls that kept it out. The challenge of their shallow admiration seemed presumptuous, almost provocative. (p. 236)

The contrasts between the touristic, colonial world and the desert are distinct and also dangerous; Felix "shudder[s]" as he watches "the cheap tourist horde go laughing, chattering past within view of its ancient, half-closed eyes. [...] These people had been wiser to choose another place for the flaunting of their tawdry insignificance" (pp. 236–237). Compared to the vitality of the desert, the "so-called 'Society' people" in Felix's hotel are one-dimensional figures, heedless and hedonistic, trapped in a narrow, petty worldview:

> Overdressed, the latest worldly novel under their arms, they strutted the narrow pavements of their tiny world [...] [T]heir essential vulgarity, their innate ugliness, seemed more than usually offensive in the grandeur of its present setting. Into the mighty sands they took the latest London scandal, gabbling it over even among the Tombs and Temples. (p. 237)

The vulgar social world spills over into the desert like a kind of pollution. Yet these people's tiny existence also provides Felix with comfort; he tells himself that "[t]he Desert has not noticed them. The Sand is not aware of their existence. How should the sea take note of rubbish that lies above its tide-line?" (p. 238). When Felix speaks of the Desert, rather than the desert, he means something numinous as well as material, and more than both. He attributes the Desert with the capacity for conscious experience, against which shallow modernity does not even register. From the Desert's perspective, human existence is *not* part of the fullness of reality; it is irrelevant against the vast scales of the Desert's existence over time.

Felix's knowledge of the desert is often given in language that encourages reading it in terms of geological deep time, contrasted to the brevity of human modernity represented by the tawdry tourist world. Felix feels the way that "one world lay upon another, but this modern layer was a shallow crust that, like the phenomenon of the 'desert-film', a mere angle of falling light could instantly obliterate" (p. 296). The modern human world is presented as a geological layer, a "shallow crust", a "film", which emphasises its fragility and thinness because human civilisation is barely visible on a geological timescale. It is also vulnerable to obliteration by the desert, which Felix later casts as "a twisting, swinging force that stormed down as though it would bend and coil the very ribs of the old stubborn hills" (p. 332). Here, the desert is a geological force, with the power to transform even the stone "ribs", the hard limestone skeleton, of the hills (let alone the fragile human world) into something malleable and elongated.

The desert's constant dynamism and power is also signalled by regular comparisons to the sea. Sand, like water, is an unstable medium that resists remaining in a fixed or permanent shape. While sand is made up of the solid remains of ground shells or rocks, it is also fluid and changeable, capable, like tides, of active movement and breaching of boundaries. Sand endows land with a "three-dimensional and turbulent materiality" that oceanic studies scholars have associated with the sea, transforming a dry environment into, nonetheless, "a world of flows, connections, liquidities, and becomings [...] a world ever on the move" (Steinberg and Peters 2015, pp. 247–248). Like the ocean and the beach, the desert's sandy surfaces are constantly remade; the tracks of *human* histories and bodies are erased. Sand, with its malleable multidimensionality, is ideal for Lady Statham's purposes as a material form for the Powers she intends to channel. It is also used by Blackwood in this way in "The Willows", in which sand temporarily bears the conical prints of other-worldly entities.

Transient as shapes in sand may be, the desert can also display the geological history of deep time in the strata of its ancient rocks, as Fig. 4.1 shows. Felix's comparisons of the desert to a sea regularly connects the present time to the distant geological past, collapsing the gulf between them. Shortly before he departs for the crucial ceremony at Wadi Hof, he is aware of the "desert tide" which "swept him down the long white street towards the enormous deeps beyond. He felt the pull of a thousand miles before him; and twice a thousand years drove at his back" (p. 309). The desert as tide is more than just metaphor. It evokes its oceanic geological past, left behind as limestone—a sedimentary rock made of minerals deriving from small and microscopic marine organisms (Collin 2004).

Fig. 4.1 1929 photogravure of Wadi Hof near Helwan, Egypt. (Photograph by Alfred Marterer. Reproduced by permission of Period Paper Historic Art LLC)

Limestone is present in much of the Sahara, produced when the region was covered by sea, as far back as the Mesozoic era. It erodes less quickly than the softer sandstone, leaving behind the fantastically shaped blocks witnessed by Felix:

> The surface limestones, resisting the erosion, block themselves ominously against the sky, while the softer sand beneath sets them on altared pedestals that define their isolation splendidly. Blunt and unconquerable, these masses now watched him pass between them. The Desert surface formed them, gave them birth. They rose, they saw, they sank down again—waves upon a sea that carried forgotten life up from the depths below. (p. 314)

Here, the passage of geological time is presented as the movement of the sea, swift and transformative, yet amid it, the limestones appear immutable: blunt, unconquerable, resisting erosion, personified as statuesque and watchful on their “altared pedestals”. Felix connects with the desert on both spiritual and geological levels; the two are blended in his phenomenological experience: "There was approach; something was drawing nearer. The Desert rose and walked beside him. For not alone these ribs of gleaming limestone contributed towards the elemental visages, but the entire hills, of which they were an outcrop, ran to assist in the formation" (p. 316). The limestone is part of the body of the desert, reflecting the inextricability of spirit and matter, or the continuity of the seen world with the unseen.

Erosion also explains the shape of Wadi Hof itself as an inverted fluvial landform, in which a valley floor becomes a ridge as the softer matter around it is eroded (Zaki et al. 2020) (see Fig. 4.1). Later in the story, Felix watches from the relative safety of the top of this ridge as Vance and Lady Statham perform their ceremony in what is now the bottom of the valley. The rocks and sand of Wadi Hof display and encapsulate the passage of time geologically, and also phenomenologically. As Felix approaches Wadi Hof, the rising moon "lifted up this tide of the Desert that drew his feet across the sand to Wadi Hof. A moment later he dipped below the ridge that buried Helouan and Nile and Pyramids from sight. He entered the ancient waters" (p. 311). Movement in space is also movement in time, connected by the extended metaphor of the moving waters that collapse temporal and spatial scale: "Time then, in an instant, flowed back behind his footsteps, obliterating every trace. And with it his mind went too. He stepped across the gulf of centuries, moving into the Past" (p. 311). Felix experiences crossing the wadi as a journey through time that erases insignificant modernity.

While Felix is thinking principally of a *human* cultural past, of "the magic of old Egypt" (p. 279), his attraction to it lies in the other forms of power and life which that magic had been able to know: "Behind [the desert's] majesty of silence lay whispers of a vanished language that once could call with power upon mighty spiritual Agencies" (p. 282). "A Descent into Egypt", too, constructs a similar atmosphere in which dormant powers are felt stirring, ancient and still yet living and mobile. In both cases, the Egyptian desert emerges as an entity whose life exceeds all human conceptions of life as a period of time-bound existence. In "Sand", beneath the apparently shifting and changing sands lies something permanent: "the eternal countenance of the Desert [...] immobile and unchanging behind these shifting veils the winds laid so carefully over it" (p. 312). In Kimberley Jackson's terms, the desert is abhistorical; its past sits not "prior to but within and against the historical" (2013, p. 130), alongside human history and without reference to it.

Felix's experience of the desert recognises the irrelevance of human history in deep time. In this sense, there is no Anthropo*cene*—anthropos has not been and will not be around long enough to shape an epoch that would show up against the background of deep time. As Felix waits, scrutinised by a Desert that "saw him" and "knew that he was there" (p. 318), he feels deep time intensely; he "could have waited thus a thousand years it seemed. [...] Caught by the sand, he stood amid eternal things" (p. 319). Felix's experience brings him into contact with the Desert's abhistorical existence and affords him a degree of understanding of what it means, in a moment of mutual recognition. Even much earlier in the story, to Felix "the Desert had no end and no beginning" (p. 248). Instead, what he experiences is a paradox as the extent of deep time as a process is compressed into the apparent stasis of sand: "Through this sand which was the wreck of countless geological ages, rushed life that was terrific and uplifting, too huge to include melancholy, too deep to betray itself in movement " (p. 248). In this way, the living qualities of the desert inhere in the slow vastness of geology, only visible when speeded up; life "rushes", but without "movement". Recognition that the Desert only *seems* lifeless because it exists on a different (geological) scale of time and motion is crucial to the story's climax as "[b]ehind the spread grey masque of apparent death lay stores of accumulated life, ready to break forth at any point" (p. 248). Such accumulations of unseen life are what Lady Statham endeavours to unleash. Here, as Felix waits, the sand of the Desert, as it does many times in the story, exists as both matter and spirit, past and present, living and dead, moving and still, collapsing the "countless" scale of geological time into a single moment of readiness: a tipping point.

Tipping points in "Sand" take a number of forms, all represented by the central motif of living sand. One such point is Felix's own sense of being on the cusp of fresh knowledge after he arrives in Helouan; he feels

> balanced on the edge of knowing unutterable things. Here was a clue that might lead him towards the hidden Egypt he had ever craved to know. An awful hand was beckoning. The sands were shifting. He saw the million eyes of the Desert watching him from beneath the level lids of centuries. (p. 269)

Knowledge, like time, is identified with an animate desert which reveals and conceals at the same time. Through its embodiment as sand, the

Desert reveals itself, as hands beckon, sands shift, eyes watch, along with its past, stretching back for "centuries" and implicating a myriad lives in its "million eyes". Later, watching the ceremony in Wadi Hof, Felix "gazed between columns, that supported the sky itself, like pillars of sand that swept across the field of vanished years. Sand poured and streamed aside, laying bare the Past" (p. 329). In these moments, sand encapsulates multiple timescales: the brief, rapid now, and the long vistas of deep time in which rock and sand are as old as the sky.

Knowledge and the past go hand in hand, threatening to overwhelm the present. Should Lady Statham's experiment succeed, another tipping point is the moment at which the power of the desert crosses the bridge into the present world. As Vance explains to Felix, "the Power, the type of life, she would waken is stupendous. And if roused enough to be attracted by the patterned symbol into which she would decoy it down, it will take actual, physical expression" (p. 307). On success of the ceremony, transformative forces normally taking place across geological timescales would take place in moments, and the consequences are not certain. On first learning what Lady Statham intends, Felix is hesitant, drawn but reluctant, just as his first experience of the encroaching desert is "appalling". But ultimately the revelations on offer are transcendent. At the height of the ceremony, as he partially glimpses "[m]agnitudes that were stupendous as the desert" with "a terrible beauty hardly of this earth" (p. 334), the "power that caught him was too great a thing for wonder or for fear; he even felt no awe. [...] The glory numbed him" (p. 335). Completion of the evocation itself fails, however, because of the actions of Richard Vance, who pushes Lady Statham into the whirling sand, which consumes her and subsides.

Real wickedness, in "Sand", is a human thing. Although Felix "feared the Desert as he might have feared the scene of some atrocious crime" (p. 299), the evil he detects is not, in the end, the powers Lady Statham is channelling; it is within Vance. Sometimes this evil feels only a "little human dread" (p. 306), too small and brief to register against the scale of Felix's experience of the desert; it is a "human horror [...] smothered for the time by the hint of unearthly revelations" (p. 298). However, this horror increases in scale until it "took the lead. [...] This abomination of a selfish human will returned to fix its terror in him" (pp. 298–299). For all the brevity and insignificance of a human life, "selfish human will" is ultimately strong enough to contend with the power of the Desert and interfere with the "revelations" of "beauty" and "glory" that accompany recognition of an other-than-human entity's existence. The reason for Vance's disruption of the ceremony and murder of his aunt is given in Mansfield's comment that "she's as rich as a pig. He follows her out here every year, waiting for her empty shoes" (p. 286), implying he is motivated by material greed. Consequently, while Lady Statham wishes Felix to draw what he sees as a way of securing a path for the desert powers, Vance wants a witness to her supposedly accidental death.

However, further motives are embedded in Vance's inability to experience the Desert on its own terms. Where Felix experiences encountering the Desert as an ecological meeting of distinct ways of being, scary but exhilarating, Vance can only view it in antagonistic terms as a binary conflict. In the absence of a "Body of Worshippers through whom it can manifest", Vance continues in his explanation to Felix of the planned experiment, the Power will "press inanimate matter into the service" (p. 307). His language emphasises not the wonder of this event but its violence: "The terrific impulse to form itself a means of expression will force all loose matter at hand towards it—sand, stones, all it can compel to yield—everything must rush into the sphere of action in which it operates" (p. 307). For Vance, there is no intelligent agency in the Desert, it is merely an "impulse". Further, he poses the event as a struggle for control rather than a gaining of mutual recognition and accommodation: the Power will "force" and "compel" the sand and stone which, passive and inanimate, "must rush" and "yield". The Power is posed as separate from the "loose matter", rather than being one and the same as sand in the way Felix experiences it.

This passage foreshadows Vance's murderous intervention at the story's climax by exposing his Cartesian ontological position in which Nature, as metonymised by the Desert, is powerful and mindless, dangerous to human existence yet inferior to it. Since Cartesian thinking has underpinned environmental degradation by conceiving Nature as resource for human use and has even led to hostility towards it, this position relates directly to Vance's material greed and explains his behaviour on both instrumental and philosophical grounds. From Vance's perspective, the encounter with the Desert is a question of domination. His consequent actions foreclose the ecological possibility of "the extension of spiritual knowledge and the widening of life" sought by Lady Statham. Her success might have threatened the "shallow crust" of modernity (which Vance seeks to protect)—or it might have saved it: "Complete rehabilitation", she adds to her earlier explanation to Felix, "might follow" (p. 277).

The story's ending is equivocal—we don't know what happens to Felix, who faints before confronting Vance, and whether the preservation of modern civilisation justifies Vance's act remains uncertain. Lady Statham herself may not have altruistic motives, and her death might be seen as punishment for her arrogant assumptions of control over inhuman powers. Felix himself, though, attains at the peak of the ceremony an ability to know reality in multiple ways at the same time:

> He saw the two things side by side: the outer immobility the senses commonly agree upon, *and* this amazing flying-out of their inner, invisible substance towards the vortex of attracting life that sucked them in. (p. 333)

Felix, at least, is able to grasp the continuity of the material and numinous worlds, leading him to a position of understanding that is at root an ecological position, ready to embrace an ontological tipping point and seek what lies beyond it. The story, then, ultimately retreats from its own vision of a spiritual tipping point that may be revelatory, or destructive, or both, but nonetheless suggests the ecological value of confronting the brink of new understandings about the position of humanity in relation to other-than-human scales and forces.

4.5 Conclusion

"Sand" collapses inhuman and human scales and modes of existence to show continuity of the human and other-than human realms at the level of geology and climate as well as at that of spirit and matter. Showing unease with the blithe materialism of early twentieth-century human colonial life at the expense of acknowledging a wider world, the story also suggests that arrogant interference in that world comes with a cost. Nevertheless, there may be value in reaching an ontological tipping point over the brink into new ways of living and being, provided it is arrived at from a position of respect. Blackwood's story models a process of reaching a radical new understanding of the world and the human place in it that can be ecologically read. Like twentieth-century crisis disciplines, Blackwood's eco-weird calls on multiple ways of knowing in responses to the prospect of ecological catastrophe.

Blackwood's weird stories return repeatedly to ontologically transformative encounters between worlds, between humans and other-than-humans, in remote liminal spaces—desert, Danube, Canadian wilderness, forest edges. In some tales, such as "The Willows", safe escape from that encounter is to be greeted with relief; in others, such as "The Man Whom The Trees Loved", union with other-than-human life is to be accepted and even embraced. In "Sand", the ultimate failure of the ceremony is ambiguous, presenting on the one hand the possibility of reaching a genuine understanding of human and planetary material and spiritual entanglement and on the other the continual puncturing of that possibility by human arrogance, greed, or narrow-mindedness.

Failure, indeed, may help explain the weird's ecological resonances as a resilient narrative mode open to trying and trying again, even in the face of unpredictability and uncertainty. The same kind of move suggests a humbler subject position on the part of weird protagonists and readers. The anti-anthropocentricity of eco-weird tales imbues them with a sense of human humility within the world and prompts readers to think for themselves, an outcome that would be reduced by a non-ambiguous ending that closed down alternatives. Through the crisis genre of the eco-weird, Blackwood's "Sand" offers readers a glimpse of how to grasp planetary scales and reconsider our place in relation to the other-than-human world, a path that may lead to greater humility and less recklessness towards it. In this sense, eco-weird uses neither ecophobia nor ecophilia, neither abjecting Nature nor valorising it. Instead, it shows itself as a mode with unique potential for confronting frightening environmental prospects such as the consequences of climate change. Neither nihilistic nor idealistic, the Weird faces horror rather than denying it, and seeks productive paths through it, even, or especially, to uncertain ends.

References

Alder, Emily. 2020. *Weird Fiction and Science at the Fin De Siècle*. Cham: Palgrave Macmillan.

Ashley, Mike. 2001. *Starlight Man: The Extraordinary Life of Algernon Blackwood*. London: Constable.

———. 2003. Algernon Blackwood and the Celtic Twilight. *Studies in Weird Fiction* 26: 2–12.

Beer, John. 1998. *Providence and Love: Studies in Wordsworth, Channing, Myers, George Eliot, and Ruskin*. Oxford: Clarendon.

Blackwood, Algernon. 1907. The Willows. In *The Listener and Other Stories*, 127–203. London: Eveleigh Nash.

———. 1910. A Psychical Invasion. In *John Silence: Physician Extraordinary*, 1–71. London: Eveleigh Nash.

———. 1911. *The Centaur*. London: Macmillan and Co.

———. 1912a. The Man Whom the Trees Loved. In *Pan's Garden*, 3–99. London: Macmillan.

———. 1912b. Sand. In *Pan's Garden*, 223–338. London: Macmillan.
———. 1914. A Descent into Egypt. In *Incredible Adventures*, 241–335. London: Macmillan.
De Cicco, Mark. 2012. 'More than Human': The Queer Occult Explorer of the Fin-de-Siècle. *Journal of the Fantastic in the Arts* 23 (1): 4–24.
Clark, Timothy. 2015. *Ecocriticism on the Edge: The Anthropocene as a Threshold Concept*. London: Bloomsbury.
Collin, Peter Hodgson. 2004. *Limestone. In Dictionary of Environment & Ecology*. 5th ed. London: Bloomsbury.
Cox, Robert. 2007. Nature's "Crisis Disciplines": Does Environmental Communication Have an Ethical Duty. *Environmental Communication* 1 (1): 5–20.
D'Odorico, Paolo, et al. 2013. Global Desertification: Drivers and Feedbacks. *Advances in Water Resources* 51: 326–344. https://doi.org/10.1016/j.advwatres.2012.01.013.
Ditlevsen, Peter, and Susanne Ditlevsen. 2023. Warning of a Forthcoming Collapse of the Atlantic Meridional Overturning Circulation. *Nature Communications* 14: 4254. https://doi.org/10.1038/s41467-023-39810-w.
Egan, Michael. 2017. Survival Science: Crisis Disciplines and the Shock of the Environment in the 1970s. *Centaurus* 59 (1–2): 26–39. https://doi.org/10.1111/1600-0498.12149.
Gallagher, Shaun. 2022. *Phenomenology*. 2nd ed. Cham: Palgrave Macmillan.
Gordon, Joan, and China Miéville. 2003. Reveling in Genre: An Interview with China Miéville. *Science Fiction Studies* 30 (3): 355–373.
Jackson, Kimberly. 2013. Non-evolutionary Degeneration in Arthur Machen's Supernatural Tales. *Victorian Literature and Culture* 41 (1): 125–135.
Lenton, Timothy M., et al. 2019. Climate Tipping Points—Too Risky to Bet Against. *Nature* 575 (7784): 592–595.
Lewis, Simon L., and Mark A. Maslin. 2018. *The Human Planet: How We Created the Anthropocene*. London Pelican Books.
Lovecraft, H.P. 1973. *Supernatural Horror in Literature*. New York: Dover. www.hplovecraft.com.
Meffe, Gary K. 2001. Crisis in a Crisis Discipline. *Conservation Biology* 15 (2): 303–304. https://doi.org/10.1046/j.1523-1739.2001.015002303.x.
Menely, Tobias, and Jesse Oak Taylor. 2017. Introduction. In *Anthropocene Reading: Literary History in Geologic Times*, ed. Tobias Menely and Jesse Oak Taylor, 1–24. University Park: Pennsylvania State University Press.
Miéville, China. 2012. On Monsters: Or, Nine or More (Monstrous) Not Cannies. *Journal of the Fantastic in the Arts* 23 (3): 377–392.
Milkoreit, Manjana, et al. 2018. Defining Tipping Points for Social-ecological Systems Scholarship—An Interdisciplinary Literature Review. *Environmental Research Letters* 13 (033005). https://doi.org/10.1088/1748-9326/aaaa75.

Miller, John M. 2020. Weird Beyond Description: Weird Fiction and the Suspicion of Scenery. *Victorian Studies* 62 (2): 244–252. https://doi.org/10.2979/victorianstudies.62.2.12.

Moore, Jason W. 2015. *Capitalism in the Web of Life: Ecology and the Accumulation of Capital.* London: Verso Books.

Myers, Frederic W.H. 1903. *Human Personality and Its Survival of Bodily Death.* Vol. 2. London: Longmans, Green & Co.

van Nes, Egbert H., et al. 2016. What Do You Mean, 'Tipping Point'. *Trends in Ecology & Evolution* 31 (12): 900–902. https://pubmed.ncbi.nlm.nih.gov/27771143.

Plumwood, Val. 2003. Decolonizing Nature: Strategies for Conservation in a Post-Colonial Era. In *William M Adams & Martin Mulligan*, ed. Decolonizing Relationships With Nature, 51–78. London: Earthscan.

———. 2008. Shadow Places and the Politics of Dwelling. *Australian Humanities Review* 44: 1–9.

Poland, Michelle. 2017. Walking with the Goat-God: Gothic Ecology in Algernon Blackwood's Pan's Garden: A Volume of Nature Stories. *Critical Survey* 29 (1): 53–69.

Punter, David. 2013. Algernon Blackwood: Nature and Spirit. In *Ecogothic*, ed. William Hughes and Andrew Smith, 44–57. Manchester: Manchester University Press.

Rockström, Johan, et al. 2009. A Safe Operating Space for Humanity. *Nature* 461 (7263): 472–475.

Rodrigues do Nascimento, Flávio. 2023. *Global Environmental Changes, Desertification and Sustainability.* Cham: Springer.

Soulé, Michael E. 1985. What Is Conservation Biology. *BioScience* 35 (11): 727–734.

Steinberg, Philip E., and Kimberley Peters. 2015. Wet Ontologies, Fluid Spaces: Giving Depth to Volume Through Oceanic Thinking. *Environment and Planning D: Society and Space* 33: 247–264. https://doi.org/10.1177/2043820619872886.

Stirling, Andy, and Mark A. Burgman. 2021. Strengthening Conservation Science as a Crisis Discipline by Addressing Challenges of Precaution, Privilege, and Individualism. *Conservation Biology* 35 (6): 1738–1746. https://pubmed.ncbi.nlm.nih.gov/34405462.

Verón, S.R., et al. 2006. Assessing Desertification. *Journal of Arid Environments* 66 (4): 751–763. https://doi.org/10.1016/j.jaridenv.2006.01.021.

Zaki, Abdallah S., et al. 2020. Inverted Channels in the Eastern Sahara—Distribution, Formation, and Interpretation to Enable Reconstruction of Paleodrainage Networks. In *Palaeohydrology: Traces, Tracks and Trails of Extreme Events*, ed. Jürgen Herget and Alessandro Fontana, 117–134. Cham: Springer.

CHAPTER 5

Weird Ecology and the Deconstruction of the Globe

Brad Tabas

5.1 Locales

This chapter is an extended meditation on tracking understood as a means of achieving heightened ecological sensitivity through a deconstruction of the global imaginary. On the traces of Baptiste Morizot's (2016, 2018, 2023) analyses of *pistage*, I follow Morizot but criticize his way of thinking about spatiality and language. For him, as for many ecological thinkers, tracking teaches us to compose discursive maps which are imagined as containing the essences of things to such an extent as to enable us to magically rediscover a more original dialogue with an animated lifeworld. For this line of thought, becoming ecological amounts to experiencing a dialectical inversion, in Elvia Wilk's words, a "merge or flip" in the relation between subject and landscape, "system," or "global" whole, whereby the subject disappears into the system and the globality of the planet comes to life as an ontologically unified subject (she calls this "death by landscape") (2022, p. 4). The weird and wildly pragmatist form of ecology presented

B. Tabas (✉)
ENSTA Bretagne, Brest, France
e-mail: brad.tabas@ensta.fr

B. H. Onishi, N. M. Bell (eds.), *The Call of the Eco-Weird in Fiction, Films, and Games*,
https://doi.org/10.1007/978-3-031-77126-2_5

here as a veering off from Morizot's track might be said to aim for the death of landscape. Not for a destruction of the global system, but for a deconstruction of the belief that one thing can contain the whole manifold of ecological relations and can thus enable a revolution liberating life from death. Weird ecology transforms the enchanted forest into a multidimensional string figure haunted with the spectral traces of portals to elsewheres. In so doing it pushes back against the missionary zeal of the globalizing Gaianism which currently dominates ecological discourse, sometimes under the name of planetary thinking. Weird ecology raises questions regarding the fantasy, here citing Jérôme Gaillardet's words, that it is possible to learn to dance to the "choreography of the habitable Earth [*la Terre habitable*]" as it seems to express itself in the "critical zone" and in light of what he describes as "scientific faith" in the power of coding "global behavior" into a "spatial synthesis" which bridges ethics and politics (2023, pp. 310, 50). It does this by suggesting that the χώρα, the place imagined as the whole significant domain of ecology in such theories, is at best an uncritical and provisional framing of what matters.

The trace or the track is the figure employed here to think of a way toward a respacing of ecology. A track is a supplement of an Other or others which enable the tracker to explore wider senses of spatiality. Weird ecology thinks of learning via tracking as a means of pragmatically developing what Roberto Unger has called "negative capabilites" (2001, p. 235). These enable us to push back our habitual and terrestrially biased constructions of place, opening us toward greater realism regarding relational phase spaces. The weird narratives which I here introduce as didactic games for trackers develop an ecology that is not grounded but *hors sol*. My aim is not to derive a system from the donation of a ground but to think in and through alienation, employing various modalities of what I have elsewhere described as "remote sensing" (2024) to signify troubling cohesions beyond the global manifold. In developing an ecological tolerance for the alien, weird ecology forms common project with what Steven Vogel has described as "thinking like a mall," priming trackers to acknowledge "the dark complexity and depth of the processes of nature that so exceed our limited ability to grasp and to control" (2015, p. 129). Weird ecology also resonates with what Timothy Morton has called "dark ecology." For the exploration of weirdness indeed plumbs "a dark pathway between causality and the aesthetic dimension, between doing and appearing" (2016, p. 5). More narrowly, though, the weirding of the ecological way presented here gestures toward an attunement with what might be called the exo-spatial. The exo-spatial is not the opposite of the globe, but rather the zone in which the pretended totality of the global reveals itself as metaphysical, other than the return to the immanence of φύσις that it pretends to be. The exo-spatial involves the planetarity of the Earth. It turns the boundaries of the global system inside out and outside in, to reveal the critical limits within hegemonic encryptions and outputs. Deconstructing the global reveals it to be an extension of various axial cosmotechnics, a product of an onto-theological metaphysics which structures space in such a way as to forbid the mixing of inside and outside, the horizontal and vertical. The ecology of the global planet becomes a sum of relational injunctions closing imaginations and deafening alternative ways of engaging with what it considers nonrational locales.[1]

Nevertheless, the alienating intrusion of exospatiality and weird spacing is embraced neither out of a belief in the liberatory potential of cosmic expansionism nor out of an affection for the vertiginous horrors of groundlessness. If anything, the motivation to confront it derives from dialectically opposed considerations. Weird ecology emerges in response to a need to think about the planetarity of the Earth in an Anthropocene that is also a Space Age. Ours is a planet in local space, not a globe whose outside is devitalized ontological negativity, a "terraqueous planet suspended in the void" (to quote Benjamin Lazier (2011, p. 606)). Beyond the grounded relation of terrestrial horizontality and lateral propinquity, the planet is in nonvoid local space, with satellites and asteroids, with other planets and even exo-planets. These are not only matters for astrobiological research and science fictional speculation. There is traffic, anthropogenic and otherwise, between zones. Weird it may seem to claim that opening our spirits to exospatiality and aliens may make us pay better attention to the Earth around us. Yet pragmatically the proposition is

[1] In referring here to axial cosmotechnics, I am referencing at once Karl Jasper's (2017) notion of the axial age as well as Yuk Hui's notion of cosmotechnics (2020a). What I have in mind here specifically are transcendental distinctions between Heaven and Earth which occurred in some cultures and not in others, and which played varying roles in the incarnations of political theology as well as the hermeneutics of historicity within these axial cultures. As Jan Assmann puts it, axiality was the age around which the "axle of world history turned," and this was so because the division between Earth and Heaven in turn played an axial role within the heuristics of interpreting historical change (2018, p. 16). Just to be clear, this is not merely a Judeo-Christian distinction. For the Chinese (to quote François Jullien), the couple "'Heaven/Earth' names the instantiated global frame" (2022, p. 56), that is, the representational manifestation of the ontological totality of what is.

valid. That is not merely a reminder of the world-revealing role that astronomy has played in the construction of modern science (Trotta 2023). It is a reminder of the well-defended heuristic utility of astrobiological studies even of seemingly lifeless planets for understanding our own Earth (Grinspoon 2009; Frank 2018). But it is above all a nudge toward acknowledging the planetarity of planet Earth, the fact that it is not ontologically a globe. If weird ecology does not aim at elaborating a post-critical cosmology, if it does not aim at landing on Earth, if it does not renounce the outside, then that is at least in part because one of its opening premises is that the outside is already (partly) inside, and that this inside outness has been one of the constitutive conditions for the globalization of the imagination which has come to believe that there is now no outside. But weird ecology does not thus aim, with Lisa Messeri (2016), to prepare future places on alien planets by projecting existing terrestrial place-imaginaries outward. Nor does it aim, as does the work of Lukáš Likavčan (2019), at comparing science fictional planets with the aim to translating what Stephenson would call a visionary "hieroglyphs" (2014) for desirable alternative planets back earthward (or elsewhere). Its target is de-globalizing our thinking on relationality, and resisting the many discourses which promise to save us from our unhealthy relation to the world precisely by embracing some sovereign relational absolute that denies activity to what it thinks of as absence.

Weird ecological thinking remains critical. It is ecology with a (weird) subject. Rather than summarily insisting that the Anthropocene condition is, in Jason Moore's words, "beyond dualism" (2015, p. 351), weird ecology embraces ontological distance but also deforms and deconstructs dualisms, eliciting reimaginings of how we structure relations between inside and out, up and down, near and far, self and other, I and they. These ongoing reconfigurations are neither a revolutionary program nor a path toward a final ontological solution. Weird ecology is an act of resistance against the siren songs of metaphysical vitalisms promising an eternal return of habitability. The happening of weirding is a pragmatic performance in self-education, a kind of testing or retraining of our faculties to respond to the exigencies of what Alexander Geppert and I have called planetization (2023). Tracking as a practice aligns with ecological *Bildung*. It further weirds humankind's already-weird status as what Evernden (1993) called "natural aliens." It liberates the exorbitant power of curiosity, deconstructs idols of globalized terrestriality in an age in which the technosphere exceeds the orb, and opens sense to the vast alien regimes and nonspecific cohesion animating our everyday. Not because that is cosmic liberation, but because it is our exorbitant responsibility.

5.2 The Space of Deconstruction

The term "planetarity," often attributed to Gayatri Spivak (2003), has been gaining popularity.[2] However, the connection between planetarity and the space around the globe, and indeed planetarity and deconstruction, remains underthought.

The connection between Derridean deconstruction and space—both in the rockets and extended substance senses of the word—which is developed in later analyses of planetarity was present from the late 1960's onward. *De la grammatologie* was published just before the Apollo moon landings, and Derrida there described his method in ways that evoke the rethinking of spatiality which was elicited by that event.[3] Deconstruction, he explained, involved attaining "a certain exteriority with respect to the totality of the logocentric epoch," adding that it was from "this point of exteriority" that "a certain deconstruction of this totality could be undertaken [*entamée*]." Like Gagarin's flight, deconstruction as a praxis opened "a path which is traced, from this orb (*orbis*) which is also orbital (*orbita*)" (1967, p. 4526). This was no single space age line. If Derrida's text often is read as being about writing, hence language, one of its central themes is nevertheless thinking of the opening toward the exterior, what he calls "the enigmatical relation of the living being with its other and with an inside and an outside: spacing [*l'espacement*]" (Jacques Derrida 1967, loc. 1819).

In so departing from surface of the orb, in so focusing on the ambiguity of the space of spacing, Derrida departed from thinking the *à venir* of philosophizing in terms of a return to Heidegger's *Erde*, situating his thought in relationship to the planetarity of the planet, the part of the planet which formed no part of the global logos, for which, as Heidegger had made clear, it was Earth and Heaven that had formed the axes of

[2] The number of works addressing the question of planetarity is too significant to properly survey. Some pertinent texts are Blake and Gilman's recent *Children of a Modest Star* (2024), the various texts by Moraru (2015), and Yuk Hui's writings on planetary thinking (2020b).

[3] There is currently no satisfying treatment of the significance of Sputnik for the understanding of space in continental thought. The closest to an adequate treatment currently available is possibly Kreienbrock (2020).

"man's poetic dwelling on this Earth," structuring the geopoetics of the Hölderlinian *Dichtung* (Heidegger 2020, p. 157).[4] In thinking the relationality of the world beyond earth and heaven, in critiquing the relationship between the voice and the phenomenon which had been Heidegger's strategy to bind poetic dwelling to the Earth via the resonances of the local "*Mundart*" (2022, p. 222), Derrida's focus on grammatology and writing followed Levinas spaceward. The latter had, in a key essay on Heidegger and Gagarin, suggested abandoning the saving power of the *Lieu*, or the pastoral place, on the motive that the Schmittian divisions between friends and enemies, autochthones and aliens, that it enabled was worse than any threats to the being of humankind posed by technology (Levinas 1984, p. 348). One might thus argue that Derridean deconstruction differentiated itself from the Heideggerian *Destruktion* of temporality in undertaking to deconstruct the orbital logic of spatial relations which formed the background for the Heideggerian notion of the place [*Ort*], and did so by insisting on an originary supplementarity which entangled techne with spacing, and so likewise led to a rethinking of timing.

Along this other path Derrida's "cosmonautical" (2008, p. 118) writing encountered satellites and supplements, Others and extraterrestrials, neither a new cosmos nor a return to place but an alternative way of performing placing and relating. Heidegger tended to imagine space-time in two opposed manners: the geometrical space of the *Satz vom Grund*, that is, the extended substance of Cartesian philosophy and the *Gestell*, and the poetic phenomenology of grounded being or dwelling which he encountered in poetic sentences. The second was the "earthly" space of terrestrial reason-giving terms within the local poetics of the German language: "*Grund* mentions the depth, e.g. the seabed [*Meeresgrund*], the valley floor [*Talgrund*], the meadow floor, a depression, low-lying land and soil; in a broader sense it means the earth, the ground [*Erdboden*])" (Heidegger 2022, p. 166). Within the pre-Sputnik place world of the *Erd* logos, at least following Heidegger, the distinction between Earth and Heaven underwrote all distinctions between reality and dreams, human and superhuman, giving shape to what he called the *Geviert*, "the inherently unique

[4] The global logos is not the logos of globalization, which was itself a counterreaction to the earliest phases of planetization. The global logos characterized the being of the world in which there was in effect no difference between the planetarity of the planet and a globe, which is to say that the space of human action could not meaningly quit the surface of the orb.

form of the unfinite relation [*Verhältnis*] of the voices of the sending [*Geschick*]" (Heidegger 1996, pp. 170–1). But after the dawn of what he called the Rocket Age, those *Grunde* could no longer suffice for thinking all that in practice existed. Earth and heaven had become a finite and no longer fundamental relation: "The rocket's orbit [*Raketenbahn*] pushes 'Earth and Heaven' into oblivion [*Vergessenheit*]. The milieu [*Wozwischen*] in which it moves is neither the one nor the other" (Heidegger 2020, p. 157). Orienting oneself in this in-between world in which humans now find themselves demands an exorbitant approach to thinking about the relationships between figures and grounds. This was not because the phenomenal world now encompassed all heaven and all earth, but because the binary diremption of previous values could not adequately encompass the new exorbitant relationalities of planetarity. For Heidegger, this was an existential catastrophe—"I was certainly horrified [*erschrocken*] when I saw the images from the moon to the Earth. We don't need an atomic bomb; the uprooting of humanity is already here" (Heidegger 2000, p. 669). What horrified Heidegger was the perspective that losing the old *Grund* might yield a total loss of meaning and a total disorientation of existence, a world in which humankind could not find ground on which it could call upon in justification of its dwelling. Yet what *De la grammatologie* illustrated was that meaning could persist and be discovered even beyond the old grounds, for poetic meaning had always implied writing, senses generated by the space of the page. For those that missed the point, *Glas* (1974), and Derrida's other experiments with columns demonstrated the exorbitant meaning of writing performatively.

For ecological thinking in a Space Age, ex-orbitant thinking imagines places and spaces outside the opposition between *Gestell* and *Grund*. It thinks of relationality beyond the logos and the orb, beyond the binary of Earth and Heaven. It thinks of satellites and astronauts and aliens. The method at work in the following is deconstructive insofar as it is exorbitant. It attempts above all to be realistic about relationality. It critically defers the satisfactions that emerge when one accepts the "holistic" mereology of global systems and thus annuls the anxieties emergent from the experience of ex-orbitance. Global system builders need limits ("in systems thinking, the first step is usually to identify your system and its boundaries" (Lenton 2016, p. 14)). In practice and by convention, those limits now generally stop at the edge of the atmosphere which is taken to the limit of the biosphere. Yet there are satellites in the orbital space beyond, and space junk as well. But that is to only talk about anthropogenic matter. Moreover, we monitor the changing planet, in Edwards words, this "vast machine," precisely from those supplements to the system which are satellites, those alien appendages which give us "global data as a matter of course" (Edwards 2010, p. 384). What is external to the system thus matters for thinking of the system. But that does not mean that the satellite zone is thus the limit. But weirdness is not a counter embrace of a cosmological or cosmicist standpoint. Exorbitance implies limits, implies inside out supplemental localizations.

5.3 Tracking

Morizot suggests we have entered a "mythic time" characterized by a "metamorphic situation" in which we can no longer give a "precise status" to things. Our linguistic and aesthetic sensibilities are in crisis. Having lost connection with the configurations of the lifeworld that we knew from myth or from the teachings of poets, our time is not only full of despair and solastalgia but also pregnant with the possibility of "strange meetings, alliances, and dialogues" (Morizot 2023, pp. 31, 38). Morizot's response to this crisis is "tracking." This is a pragmatic practice of aesthetic reorientation, as well as a foundational part of what Morizot describes as geopolitics. Tracking cultivates relations with other forms of life within common territories. These encounters and relations are supposed to produce new multispecies maps, new communities within which trackers will learn to dwell. The planetary dimension of known places that emerges out of these encounters Morizot calls the unexplored [*l'inexploré*].

One might imagine that the structure of the place world made manifest in tracking the unexplored may seem weird, that it might challenge inherited conceptions of order. Yet one has little sense of this weirdness in Morizot's writing. Indeed, sometimes it seems that the unexplored that is revealed resembles less a new world than the *Lieux* of poetic dwelling in the old figures for *Grund* evoked above.[5] This seems to be because his version of tracking does not fully open itself to the trace, but rather imposes on the tracker and the practice of tracking a telos: the return to Earth. This seems to be borrowed from a political injunction to "land on

[5] That statement is perhaps not wholly fair. Morizot draws a significant portion of his vision of place not from (as was the case with Heidegger) poets emblematic of his nation but rather from encounters with indigenous cosmologies. The conservatism manifests itself more in his interpretation of their text than in any parochialism within his objects of analysis.

Earth" and to become "Earth-bound [*terre-ester*]" voiced by Bruno Latour (2017). For Morizot, the unexplored is located not "over there [*là-bas*]," not on the "empty spaces on the map," but rather "under our feet" (2023, p. 82). But we cannot learn anything from tracking if we think that we already know where we stand and where the tracks will go. The imposition of a trajectory on a trace imposes teleological and structural expectations, an entire spatial metaphysics, which delimits horizons of interpretation which only work if that teleology is always already in preestablished harmony with that of the tracked. Word play calling the modern *sol*, soil, *hors sol*, ungrounded is not enough to point a way if the process of the composting tends elsewhere. Trackers need to be ready to think the near as far, the inside as outside, and the up as down. They need to think jumps, flights, inside outness. To a mapmaker or a thinker aiming only at fabricating ground all that may seem like madness, like utter and total disorientation. But pragmatism about performance demands a certain tolerance for weirdness. Trackers who set out to derive orientation from tracking but who begin with a biased set of expectations regarding the possible meaning of the traces do not track; they perform rituals. These liturgies may explicitly promise to serve Gaia, but they may in effect serve Cthulhu, an alien and alienating Lovecraftian monstrosity, a being that in some tales captures human minds and transports them out into the depths of space. Right intentions without right practice do not necessarily lead where one aims.

Better to focus on learning to track so as to learn from the tracking.

5.4 Space Is Part of the Land

The relational attunement that grounds the spatiality of the tracker is the instantaneous thread of relationality generated by the movement from trace to trace with the spacing of the track. That relationship can be experienced as ex-orbitant to the extent that its structuration need not solidly rest on a foundation opposing Earth and Heaven. The spatial imaginations manifested in stories told by peoples with long and successful tracking traditions often violate this axial expectation. The Lakota writer Joseph Marshall says that Native American stories are always "about the land" (2002, loc. 77). As the anthropologist Keith Basso writes of the Western Apache, the land stories not only express geographical knowledge but are also "about" how to be, about how to "organize and regulate" existence (1996, p. 53). But that can easily be misheard if we listen according to certain interpretative rules. As the Mi'kmaq astronomer Hilding Neilson has pointed out, for many indigenous peoples, "space is part of the land" (2022). The heavens are part of the earth. This is not an opposition, but a commingling. This commingling does not take the form of a sovereign exception or logical contradiction. In the beginning, Earth and Heaven were not absolutely divided. Nor are earth and heaven here manifested as they are in the Confucian *tianxia* system, in which "all under Heaven" is submitted, but only the legitimate sovereign chosen by Heaven stands in an open relation to the Heavens, not the rest of the "ten thousand things."[6] In a culture in which space is part of the land, any who are properly attuned can be ungrounded. When the anthropologist Jane Young told several Inuit about the Apollo moon landings, they replied: "We didn't know this was the first time you white people had been to the moon. Our shamans have been going for years. They go all the time" (1987, p. 272). As Mark Hollabaugh reports, there exist many accounts in which star spirits "come to the world" and sometimes "the Lakotas go to them" (2017, p. 180). Of this capacious vision of space travel, Richard Grossinger once wrote: "Given a billion years each, the Hopi, hypothetically would reach Andromeda before the American Space Program" (1970, quoted after Dias, xxxvi).

The Pawnee describe the ways of the tracker as the "Wolf-ways" (Echo-Hawk 2018, loc. 109). These are not the ways for always but for exceptional moments like being on the war path or the hunt. The pragmatic approach presented here assumes that there is not one way of ecological being; there are ways of being appropriate to each track along the trace. What does it mean to think beyond Earth and Heaven, to be on the wolf way? The ethnologist Ardy Sixkiller Clarke, who gathered stories from Native Americans regarding their encounters with aliens and UFOs, is enlightening in this regard. Ira, a Dakota Sioux who (according to other professional trackers) was "the best tracker in the Vietnam War" (2019, p. 122), met aliens when he was tracking deep behind enemy lines. He saw a blue light coming down from the heavens, followed by a boom. The zone of light radiated out and toward the tracker revealing forms which were "human-like but encased in this shimmering blue light." Ira eventually came "face to face with one of them," a being "probably nine feet tall

[6] As Jian Quin writes, for classical Confucian thought, "political power and its legitimacy originates with, or is primarily rooted in, Heaven while its recognition by the people is a secondary or derived origin" (2012, p. 183).

and muscular." From even this brief contact Ira was able to conclude that the alien "wasn't my enemy" (2019, p. 189). From this experience, Ira concluded that he and his fellow trackers now "know more about space than NASA." For he has "touched space," and through this contact, he learned that there is "life out there; something NASA is still trying to discover" (2019, p. 235). What this anecdote tells us is that Ira, while on the wolf-path, was simultaneously on a map and in a forest in Vietnam and was wandering through a liminal zone in which outer space could touch down and become part of the land. He was in a state of radical openness regarding the what and the how of the where.

Ira's is a singular tale, but based on Clarke's fieldwork it is in no way unique given his cultural context and aesthetic education. This is confirmed by the fact that she gathered over 5000 similar testimonials from what she called "Space Age Indians." The sheer quantity of witnesses suggests a certain normalness to these encounters, even if, as she notes, most also worried that they would be dismissed as "crazy Indians" (2019, loc. 1832). It is important to understand this as an Indian craziness, not as an affliction of one unbalanced individual. The Lakota scholar Vine Deloria has suggested that despite a seeming madness to these ideas there is also a grain of objective truth. These visions resonate with scientific accounts of panspermia—the idea that terrestrial life was seeded from elsewhere—as well as with the Alvarez hypothesis according to which a meteor impact caused the extinction of the dinosaurs. They also resonate with what he calls the "ancient astronauts thesis" (2023, p. 145). This is the belief that the world, not merely the biological world but also the world of technical things, exists because of an archaic visit by beings descended from elsewhere. As Deloria notes, many tribes commemorate the "genetic intrusions of the astronauts" by dividing their "people into sky and earth subdivisions" (2023, p. 149). This view is not unique to Lakota or American Indian culture. As Wiktor Stoczkowski (2022) has shown, the ancient astronauts thesis has had a long but repressed persistence within the West. Yet there is little question that including space within the land sounds far-out and crazy to those habituated to the discourse of *Grund*. The historical work of Hans Blumenberg (1996, pp. 358–376) offers some insight into how this came to be. As he points out, before Augustine "*curiositas*" was not considered a vice. But the Bishop of Hippo vilified curiosity with a specific reference to the stars above and their employments by astrologers, suggesting that attending to anything but the Book could lead to the cultivation of either "*impia superbia*" (1963, V, pp. 3–4.) if men should learn to read the stars and predict the future, or to a dangerous, because infinite, dispersion of attention given their innumerability. This led to the cultivation of forms of inwardness which stigmatized the exorbitant dangers of certain curious ways of (re)-imagining relations with the outside.

The way of the tracker, even if it is entangled with Indian and other astrological traditions, is not here being endorsed as a way toward gnostic astrological knowledge, but rather as a pragmatic practice valuable for disclosing alternate relational manifolds. Its applicability here lies not with determining and fixing the future but rather with coping with change and the fear inspired by a lack of ground. If we want, we can devitalize all of this by saying what we are really interested in is opening inner space. But the inner space/outer space distinction puts this thinking on a map, sucking out the sense of the experience. Thinking of space as part of the land more practically reminds us that the representational logics of the map and the globe are untenable in their own terms. As Arendt reminded us, looking at a map or globe is always already disentangling oneself "from all involvement in and concern with the close at hand" into what she called "world alienation" (1998, p. 251). Coming to accept that space is part of the land turns our attention to the satellite constellations polluting the night sky, traces of the fact that this alienation has now become the structure of our world. Seeing space as part of the land empowers us to criticize attempts to set up ice mines on the moon. Yet it runs contrary to any literal reading of the injunction to land on Earth, or to become *terre-esters*. Nevertheless, it forms part of the art of tracking, part of the stories and ways of being of tracker people, of cultures which are paradoxically held up by Western ecological ideologues as exemplifying how to dwell and live close to the Earth.

German philosophy describes aesthetic education as *Bildung*. Stories about weird encounters with aliens are part of Lakota and other tracker people's *Bildung*. The *Phenomenology of Spirit* was a philosophical *Bildungsroman*, a tale about the phenomenological subject discovering itself to help modern subjects to discover themselves for themselves. Ecological *Bildung* needs a different sort of tales: one that does not bring the subject back to itself, but takes it out into the weirdness of a weirding world. Weird fiction, like the stories of the trackers of the American First Nations, but unlike dogmatic ecological metaphysics of the Morizot-Latourian type, unthinks both the extended matter and the grounded and placed lifeworld. Merging space into the land may appear outrageous: an alliance with Cthulhu against Gaia. But we are no angels, no pure mediators, regardless of Latour's claims that the passion of Gaia has revealed to him a method for "exorcising" the extraterrestrialism of being modern (2022, p. 31). With respect to ecological being, it is a curious fact that the sky is sometimes beneath and inside of us.

5.5 Weird Life

Morizot imagines the nature of life in a way that directly recalls the long metaphysical tradition stemming from Plato in which the sole ex-orbitance is the sovereign sun: "we are all, we and they, heterotrophs and autotrophs, *guests* at the table of the sun, because it furnishes all of the energy here (*ici-bas*)," going on to add that "devouring the sun is the original ecological problem encountered by all living beings" (2016, p. 116). As Latour explains in a puff to Morizot's book, this planetary metaphysics of all under the sovereign solar exception is not meant (as was the case with Plato) to be "a metaphor but the foundation [*fondement*] of that which will become in the future the source of new forms of law, property relations, and sovereignty" (2016, p. 450). Yet if we might well acknowledge that Morizot aims here to inverse the other-worldliness of Platonism and thus to place emphasis on this world which will become *Grund* or *fond*, there is metaphysics within his view. This is borrowed or at least resonant with Latour's insistence that ecologists must embrace a political cosmology which "clearly distinguishes between inside and out [*le dedans et le dehors*]" based around a "new *summa divisio*" between Earth and space, without which he warns that we risk losing "the proper sense of our terms" (2021, pp. 26–7). This is, of course, an echo of Heidegger's own exorbitant fear of space, his anxiety that the extension beyond terrestrial ground would amount to a "last dark expedition of madness of the meaningless" [*letzte finstere Abfertigung des Wahnsinns in das Sinnlose*] (2022, p. 180). But in embracing this hegemonic erection within the field of relational space, in holding back curiosity for fear of vertigo, Morizot like Latour refuses to fully imagine what lies beyond the planet, and in consequence does not even think life on the home planet.

Not all living beings on earth partake in consuming sunlight. Among these are what astrobiologists call "weird life" (2013). One form of weird life are bacteria that live on the floor of the sea, surviving thanks to the heat and the minerals which emerge from deep-sea vents. Other forms of weird life include the beings of the deep hot biosphere, many of which live kilometers under the earth and consume radiation which would be fatal to more familiar kinds of beings. Much weird life, as David Toomey points out, is in practice "utterly independent of the Sun" (2013, p. 28). Many forms of weird life were discovered by astrobiologists curious about possible locations for life on other planets. They thus studied extreme terrestrial conditions that might be analogous to places such as Martian lakes, the icy moons of Jupiter, or meteorites traveling through interplanetary space. As anthropologist Stefan Helmreich has pointed out, there is something weirdly Lovecraftian about these terrestrial aliens. They are not only strangers to the communion of the sun of the more recent biosphere, but are also (like the mythic figure Cthulhu and the ancient astronauts discussed by Deloria) "ancient ones," possibly of extraterrestrial origins, who first existed during a phase of life that lacks many of the features highlighted as characteristic of habitability by contemporary Gaians. Nevertheless, there is "a cryptic but lingering record" (Helmreich 2023, p. 108) that they were once the dominant form of life on the planet, just as there are significant reasons to believe that weird life may persist on this planet even well after humans and our closest biotic kin will have gone extinct.

Weird beings are, and yet they are not, alive according to Morizot's metaphysics. This is ecologically a problem: if one wants to think about life, one needs to take in all of it, not just the life most like us. But it is also ethically and politically a problem. In more biological terms, weird life functions in Morizot's thought just as noncitizens and unregistered "aliens" function in modern states; it is left out of the symbolic order. In political terms, we call the effects of this injustice. In ecological terms, we call its effects anthropogenic system change. In less biological terms, other cultures which have curious ways of thinking more amenable to discovering and including weird life are submitted to what Fricker calls "epistemic injustice" (Fricker 2007). They may have the capacity to think about life, but according to the sovereign logic descended from the *roi soleil*, their thoughts betray life so they must be either silenced or treated as incomprehensible. Ecological philosophy betrays ecological practice when it becomes a dogmatic vitalism. In an increasingly politicized domain, these other ways of thinking planetarity risk are being branded (in Latour's Schmittian terms) enemies of Gaia. But this primarily shows the dangers of establishing a bible and a priesthood of the new *summa divisio*.

The issue here is linked with the problems of speculative metaphysics which initially powered the emergence of Kant's critical philosophy. In fleeing Platonism, but in also claiming that critical thinking has "run out of steam" (Latour 2004), actor network theory and its ecological variants run afoul of one of the basic issues with speculative ontologies: limits. If everything is connected and everything is an actor, then not only is everything on earth an agent but so also is everything in heaven. For a pragmatic theory or a critical theory, this exorbitance is not a dogmatic problem: the limits that one acknowledges are the limits of the subject, and these can be fluid, constantly in construction and deconstruction. But for the kind of speculative physics which is common among contemporary relational vitalisms, which pretend to provide a legitimate grounding for politics by describing the network of all networks, this is a problem. Just calling the object of one's concern the critical zone does not release speculative ontologies from the amphibolies of pure reason. As Elie During has pointed out, Latour and his followers have a problem with extraterrestrials, for they introduce a "reaction into our action" (2022, p. 663). But it is not merely extraterrestrials: it is the entire element of the planetarity of the planet that is the problem, that is, its location in space that matters. A speculative network without limits presumably has a politics, but those politics are not necessarily a terrestrial politics, and given the vastness of the cosmos, there is little reason why a politics of networks ought to be a terrestrial politics, as opposed to one in which all that we think is politics is, in reality, a spectacle in a microchip carried out to amuse our alien task masters. Thus, the urgent need to bring politics down to Earth and the urgent need to establish a limit between the planet and its outside are less issues of politics than they are issues of metaphysics. They are attempts to save the metaphysics of the solar logos of the orb from its own tragically self-incurred acritical contradictions.

Morizot insists we must "be done with the point of view of Sirius, the dreams of expansion that locate enchantment in space because we disenchanted the Earth" (2023, p. 101). Yet at the same time, the points of view of Sirius, or rather the nonimaginary places in extraterrestrial local space which were never identical to the point of view of a figure from fiction in the first place, offer Gaians points of view on Gaia, and are part of our historical relationship to the planet. Lynne Margulis once described Gaia as "symbiosis as seen from outer space" (2022, p. 347). Lovelock, before he died, wanted to "emulate the astronauts and fly into space to see the Earth from above" (2009, p. 1). As we have already seen, the quest to answer astrobiological questions via what Cabrol calls "analogue" (2023) missions to alien-like terrestrial environments (as well as actual robotic trips to Mars and into its deep past) has revealed dimensions of planetary life which were otherwise overlooked by traditionally grounded observers, including the fact that the climates of Mars and Venus were once like our own but have since radically degraded (Lorenz 2019). Thinking of the Heavens in the axial terms of the solar-global metaphysics makes a variegated outside into one undifferentiated block, impoverishing our sense of the nature of nature, giving birth to the figure of Sirius. Learning which perspectives and places matter does not follow from metaphysics. It emerges with curiosity. It emerges from within weird philosophies that leave critical space for spacing out life.

5.6 Weird Fiction

I want to turn now to weird fiction on the conviction that the deconstruction of the commitment to metaphysics within Western *Bildung* needs in part to be accomplished from within, and in a critical dialogue with that tradition. Lovecraft's work, and in particular the Cthulhu myth, despite (and arguably because of) his admittedly racist fear of the other, is helpful in this regard.

I have already mentioned Cthulhu multiple times, yet here I want to clarify its connection with the rupturing of the epistemic economy of the orb. Cthulhu is a "great priest" who is master of "the Great Old Ones." They are beings who "came to the young world out of the sky"—a dark planet called Yuggoth—and who "lived ages before there were any men," but who are—perhaps—gone. Yet they perhaps persist "inside the Earth and under the sea" awaiting the day when Cthulhu will return to "bring the Earth again beneath his sway," and they also appear on Earth from time to time, and these appearances set in motion the ontological space in which Lovecraft's tales take place (2008, p. 213). Reading Lovecraft's tales and Ardy Clarke's star people encounters side by side, one sees sufficient similarities to wonder whether this myth owes a debt to Amerindian lore.[7] Such a thing seems possible. In the "Call of Cthulhu," the primary informant regarding the Cthulhu cult is an "immensely aged mestizo [i.e.,

[7] Stoczkowski (2022, pp. 142–169), in his analysis of the origins of these ideas, ties them back to the gnostic tradition. Most probably, the research for origins here is a vexed question, simply to the extent that the division between Earth and Heaven is so basic to most accounts of cosmology that it is hard to imagine cosmologies that break down this "originary" but also highly political distinction not having multiple points of emergence.

half-Indian] named Castro" (2012, p. 213). But Lovecraft does not merely copy these tales: he explores the ways in which they upend and threaten the sovereign edifice of western rationality.

"The Call of Cthulhu" was published in 1926. That date is more important than it might seem: the relational manifold articulated in the Cthulhu myth can be seen as a reaction to what in Carl Schmitt's terms might be described as a "spatial revolution" (2008) associated with the dawning of the Space Age. Sputnik had not yet been launched, but the conditions of possibility for extraterrestrial travel were for the first time being put into place. Just a half-decade before, when Robert Goddard had published *A Method of Reaching Extreme Altitudes*, the first technical treatise on rocketry published on US soil, a *New York Times* reviewer had mocked him for failing to recognize that "a lack of air" in interplanetary space made space travel impossible (Neufeld 2018, p. 14). Lovecraft himself had long believed that. In a newspaper article he published in 1906, he had insisted that despite the proliferation of "imaginary journeys to the moon" within the fiction of "olden and modern times," no serious person had ever really thought space travel possible: "if any one of these authors should be asked if he thought it possible for such a voyage to be performed, a laugh would constitute a reply" (2004, p. 31). For that younger Lovecraft, the order of the Heavens was "exact, eternal, and undeviating" (2004, p. 264). The significance of this for how he had then thought about literature comes out in another newspaper piece, this one from 1915, part of his monthly column on astronomy and culture:

> Classical tales identify Aquarius with Deucalion, who with his wife Pyrra formed the only surviving pair after the deluge sent by Jupiter to punish the wickedness of mankind. This virtuous couple, who had escaped from the flood by ascending the uncovered summit of Mount Parnassus, desired at once to recreate the entire human race. Inquiring of an oracle how to accomplish this purpose, they were told to cast behind them the bones of their mother. Since Earth is the common mother of us all, and since the rocks may be said to constitute her skeleton, Deucalion and Pyrra gathered a quantity of stones, which they scattered widely about them. These stones, as soon as they reached the ground, were instantly metamorphosed to human creatures from whom we are all descended. (Howard Phillips Lovecraft 2004, p. 162)

According to the young Lovecraft, then, the classical poets understood the ordered and timeless Heavens to shed light on the origin and structure of otherwise metamorphic-seeming terrestrial being on Gaia ("Earth is the common mother of us all"). But acknowledging that humans could go to space, and likewise acknowledging that aliens might exist and could likewise come to Earth, both of which seemed to be physically entailed by Goddard's discoveries about rockets, implied a crisis in Lovecraft's conception that timeless wisdom might be derived from the stars and their poetry. It threw the very order of things off its hinges. He registered the existential crisis elicited by the historicization of the heavens in "The Color Out of Space," there describing what he had previously called eternal and undeviating as the "great temporary stars of history" (2008, p. 337).

Arguably, it is precisely the horror of this decorrelation of what in his imaginary had been the fixed foundation of the cultural edifice which he evokes as a horror so terrifying as to produce madness in the famous opening lines of "The Call of Cthulhu":

> We live on a placid island of ignorance in the midst of black seas of infinity, and it was not meant that we should voyage far. The sciences, each straining in its own direction, have hitherto harmed us little; but some day the piecing together of dissociated knowledge will open up such terrifying vistas of reality, and of our frightful position therein, that we shall either go mad from the revelation or flee from the deadly light into the peace and safety of a new dark age. (H. P. Lovecraft 2008, p. 201)

In this way, the Cthulhu myth was born out of Lovecraft's dawning acceptance that the classical order of Earth and Heaven was irreconcilable with objective reality. His later writings thus explored the "freakish curvatures in space, and of theoretical points of approach or even contact between our part of the cosmos and various other regions as distant as the farthest stars or the transgalactic gulfs themselves" (2008, p. 363) which were opened by treating the Heavens as historical as the Earth. They also registered the exorbitance of the new world as a source of horror and anxieties, signaling the transformations of established values and moral norms that this new spatial manifold implied. Lovecraft's virtue was his acknowledgment of the rupture, while the limit of his contribution to weird ecology derived from his tendency to merely bring exo-spatiality into focus as a cultural catastrophe which inseminated itself into what had been a solidly established order thanks to encounters with the horrifying alien others, and its value was further minimized by his efforts to imagine subjects who would precisely be capable of resisting the encounter with the outside, forms of life capable of touching the alien and coming back unchanged.

5.7 The Surveyor

We might call educated trackers who do not aim to learn from tracking surveyors.

This is a standard type of Lovecraftian narrator, and is also perhaps most perfectly illustrated in the narrator of "Color Out of Space." He is an actual surveyor and he has been charged with mapping an area that will soon be covered with a dam. His professional efforts soon threaten to go off the map when he discovers that the zone has been contaminated by an asteroid impact localized at the farm of Nahum Gardner. Aided in his historical research by locals, he discovers that with the asteroid came a color, one that is "almost impossible to describe; and it was only by analogy that they called it colour at all" (H. P. Lovecraft 2008, p. 171). This color ruptured the bounds of locality just as asteroid impacts ruptured classical accounts of the nature of nature.[8] It uncannily disseminates outward from the impact site, entering plants, animals, and the bodies of subjects, most especially Gardner, a figure whose name alone suggests that he has perhaps gotten too close to the natural world. One might say that just as the historicization of the Heavens created disorder within the fixed logic of matching poetic representations and their moral injunctions to an unchanging cosmos, so too did the color scramble the normal ordering and interactions of the terrestrial things in Lovecraft's tale. All efforts to explain the color, all existing causal logics and so normal science understood as the tracking of causes, are depicted as stymied. But most radical are the subjective changes wrought by the dissemination of the outside inside.

Gardner is the site of the historical outside inside. Lovecraft, speaking through the Surveyor who himself relates the words which were told to him by another, carefully documents the fate that he imagines meets those who come too close to the alienating power of the all-coloring trace: "Nothin' … nothin' … the colour … it burns … cold an' wet, but it burns … it lived in the well … I seen it … a kind of smoke … jest like the flowers last spring" (2008, p. 181). These words plot a breakdown of grammar and syntactical coherence. They are uttered immediately anterior to the death of the speaker. Intimacy with the alien, with the weird color which has landed in the depths of the well and then disseminated and

[8] On the relationship between asteroid impacts and the rethinking of the nature of nature, see Hampe's analysis of the Tunguska event (2011).

manifested even to the point of producing a weird new grammar, is thus designated as fatal, even before a dialogue which might make alternate sense of the gaps is initiated. Which is why Lovecraft's narrator, in an elaborate display of caution that is characteristic of the surveyor as subject, insists that now that the entire area has been flooded and he has returned to Boston and its civilized order, no one could "bribe [him] to drink the new city water" (2008, p. 169). This is to say that as a surveyor he retains the belief that he can acknowledge, but also stand back from, the weirding world and its alienating colors. Surveyor-being generates injunctions of this kind; it liberates and kills curiosity, never going so far as to allow the subject and language to pretend to be permanently changed by its encounter with the exorbitance of the discovery of the permeability of supposedly transcendental limits.

The posture of the surveyor is disturbingly akin to what we find in ecological metaphysics. The alterity of the physical world is evoked, but the outcome of this encounter is a return to a familiar or natural order of things, a map conveniently dimensioned to nostalgic expectations. Morizot literally insists that his book, this product of his tracking, "is not a book, it is a map" (2023, p. 11). But mapping, the practice of the surveyor, is incapable of thinking spacing ex-orbitantly. Arguably, the turn from book to map is an attempt to overcome an excessive reliance on the metaphor of the world as book present in earlier ecological thought, for example in the work of Aldo Leopold, for whom "ecological education" was a process of learning to read the "book" of natural history (Leopold 2013, pp. 43–5). One of the many problems with that metaphor is its mythologizing of the given. The world being book implies that physical traces are already legible, grammatical, and structured in such a way as to magically bring us back to nature if we only consent to pick up and read. But this is, as Rorty (2009) following Sellars (1997) has pointed out, not empirically valid knowledge but rather empirically oriented belief. Why not believe, particularly given the ecological disorder all around us, that the book of nature is written in a language like Gardner's, one that disorders our categories and deconstructs our oppositions? Speaking as a pragmatist, we would first have to try to live while using that other language to see what happened before arriving at any conclusions. The figure of the map, to the contrary, seems to pragmatically assume a certain distance with the territory, allowing one to survey the alien, to pragmatically establish certain analogies without worrying about the fact that nature literally is not a book, and things literally are not words. But even the map relies on a belief in preestablished harmony, this time located with respect to assumptions regarding the significant spatial dimensions of representability. Both the figures of the map and the territory (like the globe) code spatiality as horizontal or at best curved. They implicitly assume an empty verticality from which a surveyor, from the safety of their retreat, can judge. The exorbitance which sets the weird tale in motion is the undoing of the adequacy of the map. This is no problem if our aim—like Lovecraft's—is to subjectively shelter ourselves from acknowledging a changed world. But it is a problem if our aim is to think of the unexplored.

Wilk (2022, p. 14) differentiates between the Old and a New Weird based on the idea (in essence) that New Weird narrators do drink the water. A perfect example of an anti-surveyor is the narrator of Ruthanna Emry's New Weird story "Those Who Watch," a tale that stands in a kind of satellite dialogue with the "The Color Out of Space." From the very outset of the tale, we learn that the narrator has the color from outer space inside her. While working in a library, she picks up a book of occult literature, and suddenly she feels a sting, then "the skin swelled, red and inflamed, around a tiny spiral galaxy of indigo and scarlet flame" (2016, pp. 108–9). Of course, only those who have already read Lovecraft, and who are in a certain sense always already infected by this "bibliogenic illness" (2016, p. 117) will see this as an infection by the color from outer space. That said, she never properly identifies whatever it is as that color either, though later in the story we do learn of an occult connection between whatever has infected her and outer space: "the Tay-yug claimed that miserly gods hid favored words in the hearts of stars, making them unstable and scouring life from worlds that spun too close" (2016, p. 120). Yet as ambiguous as it may be whether what is in her is *the* color from outer space or merely some other alien resonance, what is sure is that she, unlike Gardner, does not die. She gets curious. She begins to track. She bears witness to her alienation from her environment, to the fact that she is a Northern transplant for whom the Bayou—not coincidentally one of the homes of the Cthulhu cult—is so alien and repulsive that she "can't face the swamp in person" (2016, p. 114). But at the same time, and over the course of the story and via the medium of these alien (or alienating) words, she also establishes a place, and a relation to her reality which is arguably richer than the one which could be enjoyed by any native. It might be said that the story dramatizes a double alienation which reveals much about the orb, without ever allowing for a return to ground. The tiny spiral galaxy is an emblem, not a mark of madness, death, or alienation, but an outward sign of an inward manifestation. She has turned into a remote sensor, a transmitter like a satellite which is in and out at once, beyond ground but not gravity, a device for mapping which is also, in part, off the map, not lost in space, but spaced out, and for this reason capable of offering insights into where she is.

5.8 Werewolves

The real issue with being touched with the color out of space is not death but rather the others. Borrowing a figure developed in the Blackfoot writer Stephen Graham Jones' book *Mongrels*, this is the problem of being a werewolf, a tracker, in a world dominated by surveyors. As the Pawnee writer Walter Eco-Hawk explains, in archaic times his people ceased being "Men-of-Men" and became something else, "Wolf People," beings who "respect the ways of the wolves" and themselves follow the "Skiriraa'u', or Wolf-Ways" ways (2018, loc. pp. 97–132). Among many plain Indians, the wolf ways are tied with the ways of the tracker. Interestingly, wolf-being is taken up by Morizot, who rightly celebrates becoming werewolf as a "highest form of our relational intelligence" (2016, p. 305). Yet if this is so, his writing offers precious little sense of how hard it is to hold on to this way, perhaps because he himself does not grasp the full existential significance of being a mongrel.

Mongrels is a book about growing up weird, an outsider inside. The tale tracks the life of a young not-yet werewolf. It is concerned not only with being half-wolf but also with inheriting American Indian ways of being within a society that once claimed that the only good Indian was a dead Indian, then later endeavored to pioneer forms of education which killed the Indian while saving the man. The figure of the werewolf is thus not an empty figure of the terrifying half-outsider. It is rather the fuller representation of an exo-subjectivity associated with inheriting the longer Native American tradition of exo-spatial *Bildung*. One of the most striking moments in *Mongrels* takes place in a school. The main character is "the criminal." The criminal is a child. He has not yet become a full-fledged werewolf, and he is in trouble. His aunt has been summoned because his work does not conform to the norm. Following the prompt given by the teacher—"'draw a picture'"—he has drawn a bad picture, an illegitimate mereological topology:

> The front side is the criminal on the surface of the moon. You can tell it's the moon because it has craters, and because of the stars all around, and it's because it's only a quarter full, a sliver of itself leaned back in the sky. The criminal's rocket is parked in the background. What's happening is one of the other astronauts is running up the steep top of the bowl, trying to reach the point so he can climb over, jump back to Earth. Chasing behind him is the criminal, being what he's going to be when he grows up. On the back side of the page, the one he'd had to hurry to finish, so had just red-crayoned over everything because blood floats in space, he's caught that other astronaut in his long jaws, is ripping at his stomach and somehow howling at the same time. The howl is letters filling the sky. Mostly the letter O, in a kind of kite-tail ribbon. (2016, p. 1192)

From the criminal's point of view, this is what he hopes to become. It is a wholly acceptable image of being what one is, a being who can move between earth and heaven, a being who can revel in violence, a being who longs to eat the astrocolonial other by severing the head of Whitey on the Moon. But from the teacher's point of view, this image is problematic. There is something in this Indian boy which needs to be killed—the wolf-spirit of the tracker. She is concerned that the boy's father really isn't an astronaut. She is concerned that the boy didn't understand what he was supposed to draw. Her aim is to submit the boy to the normative relationality dominant in the symbolic order before he becomes a full-fledged werewolf. She wants to kill the crazy Indian to make a surveyor who can really become a Space Man.

Part of the process of rendering being Indian socially acceptable, as Jones explained in an interview, has been ecologizing the Indian. Space Men turn Native Americans into "stewards of the forest" the kind of beings "who cry when there's a Dr. Pepper can in the creek" (Jones and Strainchamps 2021). On this view, the Native American is understood not properly as a wolf-man, but as one romantically imagines a wolf-man ought to be, a kind of elf or green angel. Angels don't eat, and if they do, they don't digest. They presumably consume what Morizot describes as "terrestrial food," something like a down-to-Earth version of ambrosia, the food of "a 'man' [*homme*] who has liberated himself from 'nature'" (2023, p. 410). The scare quotes around man and nature here indicate the vagueness with respect to how Morizot himself understands this magic flesh. In conventional terms, he certainly seems to be suggesting that terrestrial food is precisely the food that men would by nature eat, if only their understanding of these terms coincided with that of a natural being, that is, with the sort of being that the ecological Indian or the Wolf-Man who has achieved proper relationality has become. This being, to put a slightly more expansive gloss on it, would be able to consume without anxiety, since they would have found a food that resolves the central metabolic issue dogging modern societies, namely that of consumption and production. Most fundamentally, it is this which we have in mind when we speak of the Anthropocene and the need for ecology, the quest after a mode of being together in which even the simplest aspects of ordinary life, including and especially ordinary ingestion, do not produce what economists call negative externalities, that is, the unintentionally produced climate-destroying byproducts. To cite Kohei Saito, ours is an age in what was once considered progressive with regards to the "general conditions of production and reproduction" has revealed itself to be a "serious existential threat" to humans and nonhumans (2023, p. 2). Yet Morizot seems to suggest that it is not a change of all society which will yield a new metabolism, but merely learning to track down an already-given terrestrial food.

One might say that Morizot thinks of becoming a wolf as the end of the story. But Jones, traveling toward the experience of wolf-being from a wholly other orbit, begs to differ. For if *Mongrels* is about anything, it is about the problem of eating after becoming a werewolf in a world that now has a wholly other metabolism. This does not only boil down to the issue of the social acceptability of killing astronauts. It also involves attending to the fact that when one is on the track, as far out there as one needs to go, one can also consume things which are indigestible. For example: "the ragged lid of a tin can" or a "a twist tie" (both of which puncture the intestines). Because, with respect to consuming, "the wolf doesn't know any better, just knows to eat it all, and fast, and now" (2016, p. 527). For Jones—or at least for his werewolf narrator—the whole problem with the idea of *manger terrrestre* is, french fries:

> The dangerous part of the french fry is that once you have a taste for them, then, running around in a pasture one night, chasing wild boar or digging up rabbits or whatever—all honest work—you'll catch that salty scent on the air. If you still had your human mind, you'd know not to chase that scent down. You'd know better. You're not thinking like that, though. You run like smoke through the trees, over the fences, and, when you find those french fries, they're usually on a picnic table in some deserted place.

That is the human talking there, because the place really isn't deserted at all; it is just deserted from a human point of view, say located on the boundary zone between where the wolf prowls and the local MacDonalds, but more than that, there is also "a couple sitting on opposite sides of that picnic table" (2016, p. 592), or rather, was a couple, because the wolf in the werewolf likes to '*manger terrestre*,' don't you know? Which may not be a problem for the conscience of the wolf, but of course the wolf and the man both care when the hunters and their rifles show up. The point is, one can liberate oneself from 'man' and 'nature,' but doesn't mean that one has therefore gone to Heaven.

5.9 After the Picnic

Speaking pragmatically, learning how to track does not solve problems, it reveals them, often in the form of problems within solutions. Becoming a wolf-man can get you dead. Sometimes curiosity really will kill. Assassinating astronauts really is not right. If I have illustrated anything above, it is that one of the specters haunting ecological discourse is its tendency to go metaphysical by embracing overly naïve returns to *Grund*, in so doing promising salvation if we embrace ultimately empty symbols of always-already-insufficient global redemption. These soteriological strategies can yield subjects who mistake Cthulhu for Gaia and who act as Werewolves while thinking themselves Angels. Weird ecology has the modest aim of helping to educate ecologists to eat fewer fries, to kill fewer campers and astronauts, and to pay better attention to the exorbitant manifold of relations. Weird ecologists are humbly perfectionist when coping with what Cavell called the "whirl of organism" (1976, p. 52), that exorbitant and strange metabolism which is our collective form of life. They embrace a weird liberalism, not necessarily about economics, but with respect to the human relationship to the flights of our own imaginations.

References

Arendt, Hannah. 1998. *The Human Condition*. Chicago: University of Chicago Press.

Assmann, Jan. 2018. *Achsenzeit : eine Archäologie der Moderne*. München: C.H. Beck.

Augustine. 1963. *Confessions*. Translated by Rex Warner. New York: New American Library.

Basso, Keith. 1996. *Wisdom Sits in Places: Landscape and Language Among the Western Apache*. Albuquerque: University of New Mexico Press.
Blake, Jonathan, and Nils Gilman. 2024. *Children of a Modest Star: Planetary Thinking for an Age of Crises*. Palos Altos: Stanford University Press.
Blumenberg, Hans. 1996. *Die Legitimität der Neuzeit*. Frankfurt a.M: Suhrkamp.
Cabrol, Nathalie. 2023. *A l'aube de nouveaux horizons*. Paris: Seuil.
Carroll, Brian, and Hilding Neilson. 2022. *Space is Part of the Land: Indigenous Knowledges and Colonization by Light and Satellite Pollution*. Interview with Hilding Neilson.
Cavell, Stanley. 1976. *Must we mean what we say?: A book of essays*. New York: Cambridge University Press.
Clarke, Ardy Sixkiller. 2019. *Space Age Indians: Their Encounters with the Blue Men, Reptilians, and Other Star People*. New York: Anomalist Books.
Deloria, Vine. 2023. *God Is Red: A Native View of Religion*. New York: Fulcrum.
Derrida, Jacques. 1967. *De la Grammatologie*. Paris: Édtions de Minuit.
———. 1974. *Glas*. Paris: Galilée.
Derrida, Jacques, and Geoffrey Bennington. 2008. *Derrida*. Paris: Seuil.
Dias, R.V. 1970. *Inside Outer Space: New Poems of the Space Age*. Anchor Books.
During, Élie. 2022. Bruno Latour et les « Extraterrestres ». *Critique* 903–904 (8): 658–671. https://doi.org/10.3917/criti.903.0658.
Echo-Hawk, Walter. 2018. *The Sea of Grass: A Family Tale from the American Heartland*. Wheat Ridge CO: Fulcrum Publishing.
Edwards, Paul N. 2010. *A Vast Machine : Computer Models, Climate Data, and the Politics of Global Warming*. Cambridge, MA: MIT Press.
Emrys, Ruthanna. 2016. Those Who Watch. In *The Mammoth Book of Cthulhu: New Lovecraftian Fiction*, ed. Paula Guran. New York: Little, Brown.
Evernden, Neil. 1993. *The Natural Alien: Humankind and Environment*. Toronto: University of Toronto Press.
Frank, Adam. 2018. *Light of the Stars: Alien Worlds and the Fate of the Earth*. New York: W.W. Norton & Company.
Fricker, Miranda. 2007. *Epistemic Injustice: Power and the Ethics of Knowing*. Oxford: Clarendon.
Gaillardet, Jerôme. 2023. *La Terre habitable: ou l'épopée de la zone critique*. Paris: La Découverte.
Geppert, Alexander C.T., and Brad Tabas. 2023. Planetizing History. NYU Space Talks, 11 May.
Grinspoon, David. 2009. *Lonely Planets: The Natural Philosophy of Alien Life*. New York: HarperCollins.
Hampe, Michael. 2011. *Tunguska, oder, Das Ende der Natur*. München: C. Hanser.
Harman, Graham. 2012. *Weird Realism: Lovecraft and Philosophy*. London: John Hunt.

Heidegger, Martin. 1996. In *Erläuterungen zu Hölderlins Dichtung*, ed. F.W. von Herrmann. Frankfurt a.M: Klostermann.

———. 2000. *Reden und andere Zeugnisse eines Lebensweges, 1910–1976.* Frankfurt am Main: V. Klostermann.

———. 2020. *Winke I und II: (Schwarze Hefte 1957–1959).* Edited by Peter Trawny. Frankfurt am Main: Vittorio Klostermann.

———. 2022. *Denkwege.* Stuttgart: Klett-Cotta.

Helmreich, S. 2023. *Alien Ocean: Anthropological Voyages in Microbial Seas.* University of California Press.

Hollabaugh, Mark. 2017. *The Spirit and the Sky: Lakota Visions of the Cosmos.* Nebraska: Lincoln.

Hui, Yuk. 2020a. *Art and Cosmotechnics.* New York: University of Minnesota Press.

———. 2020b. For a Planetary Thinking. *e-flux Journal* 114.

Jaspers, Karl. 2017. *Vom Ursprung und Ziel der Geschichte.* Edited by K. Salamun. Berlin: Schwabe Verlag.

Jones, Stephen Graham. 2016. *Mongrels.* New York: HarperCollins.

Jones, Stephen Graham, and Anne Strainchamps. 2021. Interview: Horror author Stephen Graham Jones on What Our Monsters Say About Us, November 29, 2023, 2021, in *To the Best of Our Knowledge*, Produced by Steve Paulson, https://www.ttbook.org/interview/horror-author-stephen-graham-jones-what-our-monsters-say-about-us.

Jullien, François. 2022. *Vivre de paysage. Entre les montagnes et les eaux.* Paris: Gallimard.

Kreienbrock, Jörg. 2020. *Sich im Weltall orientieren: Philosophieren im Kosmos 1950ß1970.* Berlin: Turia + Kant.

Latour, Bruno. 2004. Why Has Critique Run out of Steam? From Matters of Fact to Matters of Concern. *Critical Inquiry* 30 (2): 225–248.

———. 2017. *Où atterrir?: comment s'orienter en politique.* Paris: La DÉcouverte.

———. 2021. *Ou suis-je?* Paris: La découverte.

———. 2022. *Qui perd la terre, perd son âme.* Paris Balland.

Lazier, Benjamin. 2011. Earthrise; or, The Globalization of the World Picture. *The American Historical Review* 116: 602–630.

Lenton, Tim. 2016. *Earth System Science : A Very Short Introduction.* Oxford Oxford University Press.

Leopold, Aldo. 2013. *A Sand County Almanac & Other Writings on Conservation and Ecology.* Edited by Curt Meine. New York: Modern Library.

Levinas, Emmanuel. 1984. *Liberté difficile.* Paris: Livre de poche.

Likavčan, Lukáš. 2019. *Introduction to Comparative Planetology.* Saint Petersbourg: Strelka.

Lorenz, Ralph. 2019. *Exploring Planetary Climate : A History of Scientific Discovery on Earth, Mars, Venus, and Titan.* Cambridge; New York, NY: Cambridge University Press.

Lovecraft, Howard Phillips. 2004. *Collected Essays: Science*. Vol. 3. Edited by S.T. Joshi. New York: Hippocampus Press.

Lovecraft, H.P. 2008. *Necronomicon : The Best Weird Tales of H.P. Lovecraft*. London: Gollancz.

Lovelock, James. 2009. *The Vanishing Face of Gaia: A Final Warning*. New York: Penguin Books.

Lovelock, James, and Lynn Margulis. 2022. *Writing Gaia: The Scientific Correspondence of James Lovelock and Lynn Margulis*. Edited by Bruce Clarke and Sebastien Dutreuil. London: Cambridge University Press.

Marshall, Joseph M.I.I.I. 2002. *The Lakota Way: Stories and Lessons for Living*. New York: Penguin.

Messeri, Lisa. 2016. *Placing Outer Space: An Earthly Ethnography of Other Worlds*. Durham: Duke University Press.

Moore, Jason W. 2015. *Capitalism in the Web of Life: Ecology and the Accumulation of Capital*. New York: Verso.

Moraru, Christian. 2015. *Reading for the Planet: Toward a Geomethodology*. Ann Arbor, Michigan: University of Michigan Press.

Morizot, Baptiste. 2016. *Les diplomates: cohabiter avec les loups sur une autre carte du vivant*. Éditions Wildproject.

———. 2018. *Sur la piste animale*. Paris: Actes Sud.

———. 2023. *L'inexploré*. Sepec: Wildproject.

Morton, Timothy. 2016. *Dark Ecology: For a Logic of Future Coexistence*. New York: Columbia University Press.

Neufeld, Michael. 2018. *Spaceflight: A Concise History*. Cambridge, MA: MIT.

Qing, Jiang. 2012. *A Confucian Constitutional Order: How China's Ancient Past Can Shape Its Political Future*. Translated by Edmund Ryden and edited by D.A. Bell and R. Fan. Princeton: Princeton University Press.

Rorty, Richard. 2009. *Philosophy and the Mirror of Nature*. Princeton: Princeton University Press.

Saito, Kohei. 2023. *Marx in the Anthropocene: Towards the Idea of Degrowth Communism*. New York: Cambridge University Press.

Schmitt, Carl. 2008. *Land und Meer: eine weltgeschichtliche Betrachtung*. Stuttgart: Klett-Cotta.

Sellars, Wilfrid. 1997. *Empiricism and the Philosophy of Mind*. Edited by Richard Rorty and Robert Brandom. Cambridge, MA: Harvard University Press.

Spivak, Gayatri Chakravorty. 2003. *Death of a Discipline*. New York: Columbia University Press.

Stephenson, Neil. 2014. Innovation Starvation. In *Hieroglyph: Stories & Visions for a Better Future*, ed. E. Finn, K. Cramer, L.M. Krauss, and N. Stephenson. New York: HarperCollins.

Stoczkowski, Wiktor. 2022. *À la recherche d'une autre Genèse: Anthropologie de l'"irrationnel"*. Paris: La Découverte.

Toomey, David. 2013. *Weird Life: The Search for Life That Is Very, Very Different from Our Own*. New York: W. W. Norton.

Trotta, Robert. 2023. *Starborn: How the Stars Made Us—And Who We Would Be Without Them*. London: John Murray Press.

Unger, Roberto. 2001. *False Necessity: Anti-necessitarian Social Theory in the Service of Radical Democracy : from Politics, a Work in Constructive Social Theory*. New York: Verso.

Vogel, Steven. 2015. *Thinking Like a Mall : Environmental Philosophy After the End of Nature*. Cambridge, MA: MIT Press.

Wilk, Elvia. 2022. *Death by Landscape*. New York: Catapult.

Young, M. Jane. 1987. "Pity the Indians of Outer Space": Native American Views of the Space Program. *Western Folklore* 46 (4): 269–279. https://doi.org/10.2307/1499889.

CHAPTER 6

Hermeneutics and the Eco-Weird

Nathan M. Bell

One aspect of the eco-weird is how it helps to highlight the otherness of the natural world and our fuzzy relationship to it—both of which can also be shown by environmental hermeneutics. To consider this I will first argue for the radical otherness of the natural world by applying Paul Ricoeur's thought on translation as a model for the limits of hermeneutics, as well as some helpful elaborations on translation by Richard Kearney. Further, I will consider Brian Treanor's ongoing work on the limits of the metaphor of the text and the irreducible otherness of the more-than-human world. After arguing that hermeneutics helps to highlight the otherness of the natural world, I will examine how this aligns with the eco-weird. Environmental hermeneutics and the eco-weird fit well together, I argue, precisely because both consider the gap between ourselves and nature and the otherness of the more-than-human world. Or, to put it another way, the eco-weird is another way to think about the otherness illuminated by hermeneutics.

Further, I will consider how we ought to respond to this otherness, considering both embodied attentiveness to nature and wonder as

N. M. Bell (✉)
Dallas College, Dallas, TX, USA
e-mail: nbell@dallascollege.edu

B. H. Onishi, N. M. Bell (eds.), *The Call of the Eco-Weird in Fiction, Films, and Games*,
https://doi.org/10.1007/978-3-031-77126-2_6

potential ways of engaging with the environment. I will also look at accepting our own limits in understanding the natural world. Last of all, I will consider translation and the eco-weird through two fictional examples, the Algernon Blackwood story "The Willows" and the Finnegan Lorcan film *Without Name*. My primary aim is to show that hermeneutics and the eco-weird are complementary to each other in consideration of environmental otherness, and further to argue for accepting a gap between ourselves and more-than-human others that can never be fully overcome.

6.1 Translation and Environmental Otherness

In her essay "Weird Reading," Eileen A. Joy considers the idea of reading for the weird. She argues that "works of literature … possess a penumbra of effects that can never be fully rationalized nor instrumentalized, and there is no one set of relations within which the whole range of any one text's possible effects can be fully plumbed or measured" (Joy 2013, p. 29). She further suggests that a weird approach to reading "is focused on (and maybe even affirms) a pluralism of being and world" and that in such an approach "it is difficult and challenging to trace the edges between self and Other, between the Real and the fabulated" (Joy 2013, pp. 30–31). A speculative reading practice, then, focuses on the multitude of meanings, the different worlds opened up by a text, and the fuzzy relations between oneself and the other.

Although Joy describes her approach as "non-hermeneutic" (Joy 2013, p. 34), I argue that her description of weird reading is not entirely at odds with certain threads of hermeneutics. In particular, the idea that we cannot exhaust the meaning of a work, the creation of multiple potential worlds, and the fuzzy relation between self and other are all commensurable with hermeneutics, at least certain variations of it. This is not to suggest that hermeneutics necessarily involves a weird reading. Rather, I contend simply that philosophical hermeneutics and the weird have some level of overlap when we consider the plurality and inexhaustibility of meaning. Or to be more specific, hermeneutics and the weird intersect at the irreducible otherness of the Other.

To consider this claim, I want to begin by considering how hermeneutics can highlight otherness, particularly in the case of more-than-human others. By hermeneutics I am referring primarily to the work of Ricoeur. For Ricoeur, hermeneutics is the philosophical view that our primary engagement with and understanding of the world is interpretation, taking the model of the text as its basis. Our understanding of the world is primarily meaningful through language, so just as we interpret literary texts, we interpret the world around us. It is important for Ricoeur that we cannot fully know or exhaust the meaning of the text. Every text, he argues, is open to various potentially valid interpretations, "open to several readings and to several constructions" (Ricoeur 2007, p. 212). However, this should not be taken to mean a sort of relativism, whereby any possible meaning is potentially valid. We can discuss and argue for different interpretations, but it is not always possible that we will come to a consensus on one particular meaning (Ricoeur 2006, p. 213). And so, there are always, to an extent, multiple possible interpretations of a text.

A key aspect of hermeneutics for Ricoeur is the relationship between interpretation and self-understanding. The way I think of and understand my self is bound up with the meanings and pre-understanding that I project onto things that I interpret. In turn, the text opens up a world in front of which I understand myself differently: "to understand is to understand oneself in front of the text" (Ricoeur 2006, p. 178). Another key aspect for Ricoeur is what he refers to as the function of distanciation. Understanding is always somewhat limited by a critical distance between myself and the text or the other. This gap is not just limiting but also productive, as this difference is what allows me to ask questions about how things could be (Ricoeur 2006, p. 142). Ricoeur argues that understanding is only complete in appropriation, "genuinely making one's *own* what was initially *alien*" (Ricoeur 2006, p. 159, emphasis in original). Appropriation is to make the text or the other meaningful to myself in ways that expand my own self-understanding. This is only possible through the distance between myself and the text; appropriation is "understanding at and through distance" (Ricoeur 2006, p. 143). I understand the text (or the other) when I understand myself and my world, in my own context, differently because of it.

Ricoeur's work considers both the potential and limits of interpretation, the extent to which we can and cannot understand the world around us and how this relates to the self, which will be fruitful for thinking about the eco-weird. This can be seen particularly in Ricoeur's thoughts on translation. Ricoeur uses translation as a model for exemplifying the limits of understanding and meaning seen in interpretation. The role of translation is to bring the reader's language and the unfamiliar language together without reducing one into the other (Ricoeur 2006, pp. 8–10). This requires "linguistic hospitality," which means recognizing the uniqueness of each language and respecting their differences (Ricoeur 2006, p. 10). We obscure meaning when we try too hard to force the other into our own language. Instead, proper translation aims at an equivalent meaning that is neither perfect nor complete. Most importantly, we never have an absolute access to the other language, or to the other. Translation is a special case of interpretation and applies to anything we attempt to meaningfully understand (Ricoeur 2006, pp. 22–25). There is always some level of alterity or strangeness—I would say some level of *weirdness*—to the text or other.

Kearney's commentary on Ricoeur's thought particularly emphasizes that translation is the attempt to understand the other or the stranger. Kearney notes that "all translation involves some aspect of dialogue between self and stranger. Dialogue means just that, *dia-legein*, welcoming the difference" (Kearney 2006, p. xvii). Translation, which again applies to any form of interpretation, is the meaningful encounter between oneself and the other, which includes recognizing our differences. Kearney's use of the word stranger is particularly interesting here, highlighting how the other is not just another entity, but is different from and strange to oneself in some way. He also states, in reference to translation, that "we are called to make our language put on the stranger's clothes at the same time as we invite the stranger to step into the fabric of our own speech." He goes on to say that "translation entails an exposure to strangeness" (Kearney 2006, pp. xvi–xvii). Kearney's discussion helps to show that translation is the encounter with the other, but in particular that the other is a stranger, highlighting their difference from oneself. Importantly, translation is the attempt to understand the stranger in a way that is meaningful to oneself without trying to fully reduce or eliminate their otherness, which entails some respect for their difference.

A last point of Ricoeur's that may be helpful is that in trying to translate or understand further, we may actually deepen misunderstanding. In translation specifically or just understanding generally, we try putting things into different words in an attempt to clarify things, though doing so may always have the opposite effect. Our further attempts to translate or understand could potentially worsen our misunderstanding—a point I think will be particularly relevant for the eco-weird (Ricoeur 2006, p. 25). We can never fully or perfectly translate another language; we will always lose some meaning when we do so. Likewise, as all understanding is through language, we can never fully understand the meaning of the other. There is always some level of alterity or strangeness—again, I would say weirdness.

I have considered at length in a previous work how we interpret more-than-human animals (Bell 2014, pp. 147–152). Ricoeur's ideas on translation, however, allow us to consider more deeply how we interpret more-than-human others and the natural environment overall. We attempt to understand the sounds, expressions, and actions of more-than-human animals in meaningful ways, much like we interpret texts and translate languages. And while plants and landscapes do not convey meaning in a similarly intentional way, as people we read meaning in the natural world, and attempt to make it meaningful to ourselves. It may be questioned to what extent nature "speaks," but in our attempt to meaningfully understand we certainly translate more-than-human others and the natural world.

We can see this most easily with animals. With companion animals such as dogs and cats, we certainly try to translate their sounds, expressions, and actions into meaningful human equivalents. We often see cats as sullen and dogs as naively happy, based on the nearest human equivalence to our expressions, though that may not be accurate. We try to understand their sounds and movements as expressing happiness, need, or warning, even when we cannot know for sure what is meant. This can be seen even further with wild animals, who are certainly more other to us than our household companions. We attempt to translate wild animals in ways that are meaningful to us. And yet, how well do we actually understand the howls and cries of wild animals? How accurate is the docility we read in the face of prey animals, and the menace we read in the face of predators? It is natural for us to read meaning in the sounds and expressions of wild animals, but we have to recall the differences between their "languages" and ours. This is where the model of translation is helpful, in noting the limits of our ability to understand more-than-human animals. There is a tension between wanting to genuinely understand the other, and forcefully reducing the other into familiar terms and meanings. This is where Ricoeur's concept of linguistic hospitality is helpful, as we have to accept at best a limited equivalence between an animal's expressions and our attempt to meaningfully understand them.[1]

[1] I discuss the translation of more-than-human animals more in-depth in my chapter "Translating Nature: Hermeneutics, Otherness, and the Limits of Environmental Understanding" (Bell, forthcoming).

This view of translation even applies, I would argue, to plants, landscapes, and ecosystems. Plants convey biological health through their visible appearance, but aside from that to what extent do they convey meaning? Plants do not move or make sound in the ways that animals do; they do not give us the same kind of intentional signs to translate. Despite this, we do attempt to meaningfully understand and to read meaning into the existence of plants. We certainly view plants as flourishing or diminishing, and may even project joy or sorrow, at least metaphorically, onto their appearance. That said, we need to avoid projecting too much or reducing them into our language.

Similarly, we often attempt to find meaning in landscapes or biotic communities. We might take the different sights, sounds, and scents of a particular forest, for example, as making up a sort of language. As such, we may intuitively try to understand, or translate, the forest as a whole—or likewise with any other biotic community, natural area, or landscape. Again, we have to be aware of the tension between our attempt to understand and reducing the other(s) too much into our own language. To take the example of a forest again, we can certainly read meaning in the combination of sights and sounds—for example, when a combination of sounds indicates an active community of forest life, or when the absence of sound indicates a threat or disturbance. On the other hand, we should not force meanings that are not there; an extreme example of this may be when we think of the forest as a harmonious community as a way to romanticize it.

Again, translation may be a helpful way to think about how we attempt to see meaning in animals, plants, and biotic communities. The relevance of a hermeneutic view of translation, I argue, is showing the limits of understanding ecological others. With two human languages, we can never perfectly convey the one language into another without some loss of nuance of meaning. Even more so with animals and wild places, our translation of them is limited and will always lose or distort some meanings. I do not mean to suggest we should not try to understand the more-than-human world; we can only understand things by putting them into terms meaningful to us. However, we will never perfectly or fully understand environmental others. It is also important to remember that further translation can make misunderstandings worse. Trying harder to conceptualize or understand ecological others may actually make us understand them less or increase the distance between us.

I want to touch briefly again on Kearney's specific use of "stranger" and "strange" in relation to translation (Kearney 2006, pp. xvii–xviii). More-than-human others are particularly strange to us. Wild animals have an appearance that seems familiar and different at the same time—many of them somewhat resemble the cats and dogs that are companions to many of us. At the same time there is something radically different to them, and their expressions and actions are often unsettling in their difference. Plants are even more so, as we recognize them as living things, but living things that are so fundamentally different from ourselves. There is something strange about more-than-human others. Kearney clearly notes "the common experience of tension and suffering which the translator undergoes as he/she check the impulse to reduce the otherness of the other, thereby submitting alien meaning into one's own scheme of things" (Kearney 2006, pp. xv–xvi). Our attempts to understand more-than-human others are not only limited but may be disruptive or disorienting to our sense of self.

While we can and should attempt to translate the more-than-human world in ways that are meaningful to us, we always need to be aware of the risk of reducing ecological others by trying too hard to put them into our language. This, I argue, helps to highlight the *otherness* of more-than-human others. We can never *fully* understand ecological others or the natural world; we cannot translate the full extent of their existence and meaning into our own term. Whether we are considering wild animals, landscapes, or whole ecosystems, there is always an insurmountable otherness—or *weirdness*—to the more-than-human world.[2] And accepting this weirdness as something we cannot fully reduce creates tension. Below I will consider how the eco-weird is a fruitful way to think about the otherness of nature that is highlighted by hermeneutics.

First, however, I want to look at Treanor's view on hermeneutics and the otherness of the natural world. Treanor considers the limits of our ability to understand more-than-human others based on the difference between human language and the expressions of the more-than-human world. Treanor argues that, while animals at least certainly communicate, they do not engage in the rich creation of meaning that is "the heart of human language" (Treanor 2023, pp. 16–17). Our interpretation of more-than-human others is limited due to the difference between our languages and theirs. Treanor likewise touches on translation and the risk of

[2] I explore some of these themes further in "Translating Nature" (Bell, forthcoming).

distorting the meaning of ecological others. He argues that "because language is distinctive to humans, when we apply the linguistic paradigm to 'capture' and 'translate' the meaning of the more-than-human world, the danger of distortion is particularly acute" (Treanor 2023, pp. 17–18). When we attempt to translate ecological others, we risk forcing our own meaning onto and over them.

As such, Treanor argues that we need to accept an unbridgeable distance between ourselves and the natural world. There is something about more-than-human others and the natural world which is beyond our full grasp. Treanor goes on to say that "the strangeness of the non-human world eludes us in part *precisely because of its non-linguistic otherness*" (Treanor 2023, p. 16, emphasis in the original). He further argues that "in the case of nature, we must remain attentive to both our kinship with it and its profound otherness and wildness, its irreducible mystery" (Treanor 2023, p. 18). We could say that the *weirdness* of the nonhuman world eludes us because of its nonlinguistic otherness. Hermeneutics in the vein of thinkers like Ricoeur and Kearney helps to highlight the otherness of the more-than-human world, as shown in work by myself and Treanor. Now I want to consider how the eco-weird can be a fruitful way to consider this otherness.

6.2 Hermeneutics and the Eco-Weird

My aim here is to consider how environmental hermeneutics and the eco-weird may complement each other, both in thinking about the otherness of the more than-human world and in how we address that otherness. According to Brian Onishi, the eco-weird "focuses on the uncanniness of nature and environmental dangers" (Onishi 2020b, p. 158). There is something about the natural world, whether we look at animals, plants, or ecosystems, that is strange to us. The more-than-human world is something other, something which at times resists our tendencies to neat categorizations. In another work he links the eco-weird to "entanglement thinking," which "requires that we embrace the fuzzy borders between objects, entities, species, and ecosystems" (Onishi 2022, p. 39). Again, we tend to want to put the natural world into neat categories of thinking, particularly regarding individual others, species, and ecosystems. That makes it easy to meaningfully consider ecological others, but does doing so obscure their actual entangled existence? Here we can see the point where hermeneutics and the eco-weird can converge. Hermeneutic philosophy raises the concern that in our attempts to understand things, we force our own meanings over them—this risk can be seen particularly in hermeneutic thoughts on translation. And as noted above, Treanor clearly shows how this relates to considering more-than-human others and the risk of distorting meaning. As Onishi argues, the eco-weird tries to consider the ways that the existence of things, particularly in nature, may resist our usual neat boundaries. Both environmental hermeneutics (at least in certain thinkers) and the eco-weird consider how our classifications of the more-than-human world, and our desire to understand fully, clashes with the real existence of that world.

There is also the linguistic gap between ourselves and more-than-human others. Onishi focuses particularly on plant life: "Perhaps the strangest and most unsettling part about our relationship with plants is that we seemingly do not have a shared language or means of communication. We lack access to plant life" (Onishi 2020b, p. 160). He goes on to say that, due to this, there is something *alien* about plants. Again, we can see this in what Kearney says about translation and the stranger, as well as both Treanor's thoughts and my own on our limited ability to understand more-than-human others. Whether we are considering plants or wild animals, there is a limit to our ability to communicate, and correspondingly a limit in our ability to understand what it is like for them. This is a further point on which environmental hermeneutics and the eco-weird converge. When we attempt to listen to the natural world (either metaphorically or literally), we are listening to others whose communication is radically different from our own. Pointing out the difference in communication, Treanor claims that "in one sense, human language exists on a continuum with other forms of animal—and, indeed, plant—communication," and yet "sometimes a difference in degree is so extreme as to constitute a difference in kind" (Treanor 2023, p. 16). Both hermeneutics and the eco-weird address the linguistic and epistemological gap between ourselves and the more-than-human world.

This gap can also be seen in our own inability to fully conceptualize the otherness of ecological others. Onishi argues that the eco-weird highlights how some aspects of the natural world, some layers of its being, resist what we can put into meaningful thought. He states that "the idea that there is a world outside of our observation, outside of *our very thought* is already weird. It creates a feeling of uncanniness specifically because it *cannot be thought*. It is some lost remainder, a haunting specter that never crosses into our frame of reality" (Onishi 2020b, p. 168). Likewise, the hermeneutic model of translation applies even to our own thinking, our attempts to understand the other in a way that is meaningful to ourselves. As Kearney notes, "for Ricoeur, the task of outer translations finds echoes in the work of inner translation." He goes on to say that "the answer to the question '*who* are you?', always entails a translation between the self and others both within the self and outside the self" (Kearney 2006, p. xix). This is important because it reminds us that translation can be internal. It is not limited to the outwardly written or spoken word, but it applies to our ability to make things meaningful to ourselves. The limits of translation apply just as much to the way that I can make more-than-human others meaningful to myself. This again relates to the eco-weird, and a focus on the aspects of the other that go beyond what I can meaningfully comprehend.

For hermeneutics, our interpretation and translation always relate back to self-understanding. Whether we are referring to the interpretation of a text or the translation of the other, it always comes back to how I understand myself in front of the other (Ricoeur 2007, pp. 177–181; Kearney 2006, pp. xix–xx). This is helpful for considering the weird as well. The eco-weird focuses on the disruptive aspects of the otherness of the more-than-human world, the ways in which the strangeness of the natural world is displacing and disorienting. This makes sense from the perspectives of hermeneutics; perhaps the weirdness of nature is so disruptive because it displaces my own sense of self. If the self is defined through the detour of the other, then who am I in the face of the other who is weird and unknown? Who am I in a world whose margins are beyond my very grasp? The weirdness of the world shakes up the foundation on which we place the self.

Last of all, we can see a connection between the eco-weird and hermeneutics insofar as each of them focuses on possibilities; the denial of an absolute known does not close off meaning but creates openings. In Ricoeur's hermeneutic philosophy, he notes that "if it is true that there is always more than one way of construing a text, it is not true that all interpretations are equal … The text is a limited field of possible constructions" (Ricoeur 2007, p. 213). This also applies to our interpretation of the natural world. It is not a pure relativism where anything goes, but there are always multiple potentially valid ways to interpret the text, including our understanding of the more-than-human world. This can also be seen in Ricoeur's philosophy of translation. The only way to critique a translation is to offer an alternative one, to consider another layer of meaning

(Ricoeur 2006, p. 22). Hermeneutics, at least in Ricoeur's work, opens us up to a plurality of potential meanings. Turning to the eco-weird, Onishi argues that "the weird irruption of nature has offered an aporia to an unfamiliar world where new possibilities emerge" (Onishi 2020b, p. 172). The eco-weird is, like a hermeneutic consideration of the natural world, not about shutting out meaning but being open to further possibilities in the absence of an absolute meaning.

And so, we can see the ways that environmental hermeneutics and the eco-weird may be seen as complementary. Both consider the radical otherness of the more-than-human world, and the limits of our attempts to have a firm grasp of the meaning of such others. Both consider how this otherness is disruptive, and how this opens new potentialities for meaning. The remaining question is how we ought to respond to the weirdness of the natural world.

In his work on the otherness of the more-than-human world, Treanor draws on the carnal hermeneutics of Kearney to formulate a way to approach this otherness. In order to empathize with nonhuman others, Treanor argues, we need to be open to nonlinguistic expression and be able to listen deeply in new ways. Part of this deep listening means holding back our urge to immediately put everything into firm categories. We also ought to remember that like more-than-human others we are physically embodied beings. We should listen to the material expressions of the natural world and focus on bodily engagement. This is not meant as a replacement for our usual linguistic understanding, but as a supplement to it. What Treanor suggests is to focus on different ways of engagement with the more-than-human world, as each way will obscure some meanings while highlighting others (Treanor 2023, pp. 21–25). Kearney also briefly considers an embodied connection to the world, noting how we can engage reciprocally through what he terms "double sensation" (Kearney 2023, p. 246). Importantly, we often reduce more-than-human others into being *for us*. Instead, we can focus on letting them *be*. Both Treanor and Kearney discuss engaging with the natural world without reducing more-than-human others into our meaning, and instead letting them be in their otherness.

So how does this response resonate with the eco-weird? Treanor and Kearney both focus somewhat on more positive aspects like empathy or connection. But the idea of an embodied engagement could also include the aspects of the world that are threatening or frightening. The bodily feeling of unease at the sound of a deep howl, or the chill we feel when the woods suddenly seem ominous or threatening, are also embodied engagements with the natural world. This applies to the visceral revulsion we might feel at the sight of ecological destruction or pollution as well. A bodily engagement with the more-than-human world will include a range of reactions and sensations, from joy at the beauty of nature, to the unease or fright when the wilderness seems threatening. An embodied engagement includes all the ways that the natural world disorients or frightens us, all the ways that it is weird or strange.

I also want to briefly address wonder as well. Onishi considers wonder as a way to respond to unsettling or uncomfortable encounters with the natural world. After all, he argues, wonder is not merely positive but can also be unsettling or disorienting. In particular, Onishi notes that wonder can dislodge us from our already-held ways of thinking and open the space for us to consider things in new ways (Onishi 2020a, pp. 213–218). This in turn relates back to a key aspect of hermeneutic thought, which is being open to the potential for new meanings and open to various interpretations and translations. Likewise, wonder may relate to Treanor's idea of a deep listening. Onishi argues that "wonder, if nothing else, demands an intensity of attention and most often leads to a re-enchantment of the object of wonder" (Onishi 2022, p. 38). A deep listening or attentiveness, combined with the suspension of our tendency toward immediate categorization, can open us to see things in new ways. In particular, it can allow us to view the weird aspects of the more-than-human world in ways that reveal what was previously hidden. More importantly, it allows us to consider the value and importance of something that we may not fully understand.

There are various ways, I argue, in which a hermeneutic consideration of environmental otherness and the eco-weird as a concept overlap. This is not to suggest that one reduces into the other. I would hesitate to say that hermeneutics is necessarily weird; rather just that the eco-weird is another way to think about the gaps and otherness that are revealed by hermeneutics. I would also hesitate to say that the eco-weird is necessarily hermeneutic as such. However, the two overlap in allowing us to think about the radical otherness of nature, the plurality of meaning, and the gap in understanding between ourselves and more-than-human others. And both, I suggest, argue that instead of reaching for an absolute grasp of the natural world, we ought to accept the limits of our understanding and respect the irreducible otherness of nature.

6.3 Translation and the Eco-Weird in Fiction

Last of all, I want to consider a couple of fictional examples, both in literature and in film, to think about translation, the eco-weird, and the otherness of the natural world. As shown above, for Ricoeur the point of literature is ultimately to help us understand ourselves and our world differently. He argues that "through fiction and poetry, new possibilities of being-in-the-world are opened up within everyday reality. Fiction and poetry intend being, not under the modality of being-given, but under the modality of power-to-be. Everyday reality is thereby metamorphised by what could be called the imaginative variations which literature carries out on the real" (Ricoeur 2007, p. 142). Fiction has the power to show us truths about ourselves and the world in which we live, which can be seen in considering literary examples of the eco-weird.

First, I want to consider Algernon Blackwood's short story "The Willows." The story follows an unnamed narrator and his traveling companion, referred to only as "the Swede," as they traverse down the Danube River by canoe. The pair stop to camp for the evening at a small island, merely a high point of sand and willow bushes that may disappear as the river rises. Early on the narrator is struck by the presence of the willows, which is not to say their mere existence but rather the extent to which they seem to him more as entities than a mere part of the backdrop. He states: "These willows never attain to the dignity of trees; they have no rigid trunks; they remain humble bushes, with rounded tops and soft outline, swaying on slender stems that answer to the least pressure of the wind; supple as grasses, and so continually shifting that they somehow give the impression that the entire plain is moving and alive" (Blackwood 2011, p. 1). Of course, the willows as plants are living things, but the landscape, the willows in particular, strike the narrator as being *alive* in a way that is seemingly agential, more akin to people perhaps than how we usually think of landscapes. As the pair set up their camp, they are horrified to see a dead body floating in the river, only to laugh with relief as they realize it is merely an otter.

As the pair make camp for the night, the narrator is increasingly struck with a sense of unease. He links this unease with the sound and sway of the strong wind through the willows. He describes the sound of the willows in terms similar to language: "But the willows especially; for ever they went on chattering and talking among themselves, laughing a little, shrilly crying out, sometimes sighing—but what it was they made so much to-do about belonged to the secret life of the great plain they inhabited" (Blackwood 2011, pp. 6–7). And he describes the movement of the willows more and more in terms of moving agent rather than a mere natural phenomenon. The narrator states that the willows "moved on their own will as though alive, and they touched, by some incalculable method, my own keen sense of the horrible." In his growing unease he begins to feel as though he and his companion are intruders in an "alien" world, humans in a part of nature where they are not welcome: "The willows were against us" (Blackwood 2011, p. 7). Though he attempts to shake away these thoughts as the product of his imagination, the narrator cannot help but see the landscape on the small island as a hostile one. The willows are not just mere plants, but living agents that are a threat to him.

The longer the pair stay on the island, the weirder things become. The narrator sees strange figures arising from the willows in the night, which he cannot dismiss as mere imagination despite his best efforts. In the morning, the narrator is sure that the willows are closer to the tent than they had been previously. He increasingly thinks of the willows as hostile agents: "There was a suggestion here of personal agency, of deliberate intention, of aggressive hostility" (Blackwood 2011, p. 10). The narrator and the Swede find that their canoe is damaged, trapping them on the island another day as they make repairs. They also discover that one of their oars and some of their food has been taken in the night as well. And they find small holes in the sand, which they later describe as "sand-funnels" (Blackwood 2011, 15). The narrator repeatedly tries to offer natural explanations for this which are dismissed by the Swede, who has by now revealed that he shares the narrator's view of the willows' agency and hostility.

As the second day goes on into night, the pair begins to hear a strange sound, seeming to come from the willows at times, and other times from above them. The narrator's companion argues that they are in a convergence between their world and another, and that the forces of the other world are a threat to them, arguing that the sound is "in the willows. It's the willows themselves humming, because here the willows have been made symbols of the forces that are against us" (Blackwood 2011, p. 19). The pair discuss how the forces against them seek not physical harm, but mental harm. The risk they face, or believe they face, is not a physical death but a loss of consciousness—a loss of self. The companion argues that only the sacrifice of a victim will save them. While fleeing from a shadowy figure, the two men fall. The companion believes this fall saves them, as the narrator's pain and the companion's loss of consciousness breaks them from the mental grasp of their foe. At once the noise stops and the threat seems gone. The pair go to sleep, intending to leave once it is light out again. In the morning the pair find a dead body among the willows, marked in the face and chest with the same funnels as the sand. As the body floats away down the river, the narrator notes how much it looks in the dim light like an otter floating away.

The story of "The Willows" shows both the idea of translating nature and the otherness of nature found in the eco-weird. The narrator clearly views the sound and movement of the willows as a type of language; again, he suggests that the willows are "talking among themselves, laughing a little, shrilly crying out, sometimes sighing" (Blackwood 2011, pp. 6–7). The willows are trying, or so he seems to suggest, to convey something. And yet, though he views the sound of the wind in the willows as a form of meaningful expression, that exact meaning is incomprehensible to him. He cannot understand the expression of the willows, other than a vague sense of horror that it instills in him. He describes the sounds of the landscape as having their own hidden intentional meaning, something which he is not meant to know or understand.

The inability of the narrator to understand nature also shows themes of the eco-weird. The willows are entities that are alive and convey meaning, not merely a part of the world but more distinctly a type of Other that, like in much of weird fiction, poses as an otherworldly threat. But this Other is shown to them not in the form of some monstrous entity, but in nature itself; the companion describes the willows as the symbols of the forces against them. And it is in this threat that we can see an overlap between hermeneutics and the eco-weird. As mentioned above, the threat against them is not physical but ultimately mental, the loss of their own conscious self, and clearly centers the idea of the relationship between the self and the other. In this case, however, it is not merely the understanding of the self in front of the other, but the risk of the annihilation of the self. The eco-weird presents the natural world distinctly as an Other, and particularly as an Other that confronts the meaning of the self.

For a second fiction example, I turn to Lorcan Finnegan's film *Without Name*. The movie focuses on Eric, a land surveyor hired to map a forest somewhere outside of Dublin, Ireland. Early in the move we see a small plant growing through a crack in the concrete, which would not seem particularly important if not for the camera's focus on it, Eric's lingering gaze, and the ominous music that accompanies this scene. The company that hired Eric has put him up in a small cabin where he finds a handwritten book, "Knowledge of Trees," written by the cabin's previous occupant, Devoy. As Eric begins to survey the woods, mysterious things start to occur. His equipment malfunctions, and he often finds his equipment having been disturbed the next day. Later in the movie, an entire day's worth of data has randomly disappeared from his computer. There are also times where Eric seems to lose himself as the light shifts strangely in the trees, and we get brief glimpses of a dark, shadowy, earthy looking figure in the woods—a parallel to the green man, perhaps.

One night in the nearby village, Eric learns more about the woods and Devoy from Gus, a traveler who has been camping just outside the woods for a few years. The woods are "unmapped, unnamed," though the locals refer to the area as *Gan Ainm*, literally "without name." There is also a local superstition that the woods are "fairly fields" where human wanderers may be lost forever if they do not enact certain rituals (*Without Name* 2016). Eric is joined by his assistant and affair partner Olivia, who experiences the same loss of time and strange shifting light out in the woods that Eric did earlier. Later, Gus tells Eric and Olivia more about Devoy, the previous resident of the cabin. Devoy believed that the trees were sentient. He attempted to photograph the aura of damaged plants, and he described the swaying of trees as a form of sign language. Devoy also used hallucinogenic tea made from local mushrooms, which he thought provided a "guide to Earth's language" (*Without Name* 2016). He was ultimately found naked, hypothermic, and catatonic out in the woods.

Later, Eric looks at Devoy's photographs of plant auras and listens as Olivia reads from Devoy's book:

> Is there grammar in all the greenery? Is there syntax in the spawning? Are leaves linguistic? Is there something worth reading in the scripts of bent trunks and stems, in crooked branches and petals? I know there is. I've come to understand there's a book of nature, and all I have to do is find out how to open it, turn the pages, and read the words. (*Without Name* 2016)

Olivia scoffs at the writings and attributes them to Devoy's use of hallucinogens. But Eric begins to think that Devoy was right. Later we see a face in the moss on a tree, and Eric hears heavy human breathing, not his own, as he looks around the empty woods and flashes back to the plant growing through concrete. We see further glimpses of the figure in the woods, and at night we see the shadows of the trees reaching down toward Eric as he sleeps.

After making hallucinogenic tea from local mushrooms and using Devoy's own recipe, Eric hides in the woods as his family and friends search for him. He confronts the figure in the woods, stumbling as he attempts to form words, and finally screams "communicate!" The figure is revealed to be Devoy, who is at the same time laying in a Dublin hospital unable to move or speak. The two fight, with Devoy strangling Eric. Eric is found catatonic in the woods and is taken into a Dublin hospital as Devoy, in the same hospital, becomes fully conscious for the first time in years. The film ends with Eric, or the other Eric, standing among the trees, silhouetted in moonlight, the new shadowy figure in the woods (*Without Name* 2016).

The movie has clear parallels to both the idea of translating nature and the idea of the eco-weird. Eric begins the movie doing survey work, trying to neatly map out the wooded area—to reduce everything down to scientific data. He argues with Olivia, on two separate occasions, that they are just collecting data, and what their employer does with that is not their responsibility. He later, however, begins to think that the woods are trying to tell him something, something that cannot be measured or mapped out. As mentioned above, he drinks hallucinogenic tea in an attempt to open himself to what the plants are trying to say. This could be read as an attempt to interpret or to translate nature. He also seems to begin to lose his sense of self as he opens himself more and more to meaning of plant others, culminating in what could be seen as a separation of his physical and mental self. Additionally, Devoy's writing, as the most forceful example, clearly discusses the idea of reading meaning in the trees.

The movie also shows that the natural world is *weird*, which is to say beyond our usual neat categories. The film uses camera zooms and ominous music to give weight to Eric snapping off a tree branch and forcefully stabbing the tips of his survey tripod into the ground, in addition to the aforementioned plant growing through concrete—the plants and trees and the earth itself seem to struggle and feel pain, which Eric and other people are, at least initially, entirely oblivious to. There are also the attempts first by Devoy and then by Eric to understand the meaning of the natural world, to "open [the book of nature], turn the pages, and read the words" (*Without Name* 2016). And yet despite their efforts, we are never given a clear reason to think that they fully understand what the plants and trees are trying to tell them. The movie ends with Eric, or perhaps Eric's spirit, standing among the trees, separated from his other self that belongs to the human world. And yet, we do not get the sense that Eric fully belongs there. Our last clear shot of Eric's face shows him looking lost and confused, a stranger among the trees.

A further analysis of *Without Name* is given by Dawn Keetley. The real narrative power of the film, Keetley argues, is in its ability to foreground nature, to center what is normally just the background (Keetley 2016). Further, the film shows the dichotomy between the human world and the natural world. Scenes of Eric's clearly unhappy urban life "are sharply contrasted with the lushness and vibrant greenness of the woods he comes to survey, a place where another life than our own—plant life—vies with human life for dominance." Out in the woods the plants are a type of other, in their own domain where humans do not belong: "the trees swaying and cracking in the wind do indeed seem to become a palpable sentient force, and one can feel their indifference toward anything human" (Keetley 2016). Keetley compares *Without Name* to Blackwood's "The Willows," noting how both stories foreground the natural world. She argues that "Because they are really *seeing* nature for the first time, the characters in Blackwood's story experience it as fully alien, and it's only *after* experiencing nature as uncanny that they see strange figures looming darkly out of the willows—figures that seems an embodiment of nature's own eerie power" (Keetley 2016). Though Keetley does not discuss translation or the eco-weird in her analysis, her comments are helpful for thinking about the intersection of the two. In each case, the story presents more-than-human others as entities that are not just living things but alive in a way similar to humans, as having their own domain that exists outside of human control and meaning. I would suggest that the eco-weird, as seen in both "The Willows" and *Without Name*, presents the natural world as being a type of self to which *we* are the other. And the other-self of nature is threatening to us not just because of its vastness, but because it has its own meaning that will always be beyond our full grasp.

In conclusion, then, I have argued that environmental hermeneutics and the eco-weird can be complementary ways to consider the limits in our understanding of the environment, the gap between ourselves and more-than-human others, and the radical otherness of nature. While there are various ways to respond to environmental otherness, including things like deep attentiveness and embracing wonder, above all else we need to accept the limits of our understanding and be open to a plurality of potential meanings. We can also see the overlap of hermeneutic translation and the eco-weird in the stories of "The Willows" and *Without Name*, where more-than-human others are other selves that have their own meaning and existence beyond the scope and control of human understanding.

References

Bell, Nathan M. 2014. Environmental Hermeneutics with and for Others: Ricoeur's Ethics and the Ecological Self. In *Interpreting Nature: The Emerging Field of Environmental Hermeneutics*, ed. Forrest Clingerman, Brian Treanor, Martin Drenthen, and David Utsler, 141–159. New York: Fordham University Press.

———. forthcoming. Translating Nature: Hermeneutics, Otherness, and the Limits of Environmental Understanding. In *Environmental Hermeneutics in the Anthropocene: Nature and the Conflict of Interpretations*, ed. Forrest Clingerman, Brian Treanor, and David Utsler. New York: Routledge.

Blackwood, Algernon. 2011. The Willows. *Blackwood Stories*. https://algernon-blackwood.org/Z-files/Willows.pdf. Accessed July 20, 2024.

Joy, Elaine A. 2013. Weird Reading. *Speculations* IV: 28–34.

Kearney, Richard. 2006. Introduction: Ricoeur's Philosophy of Translation. In *On Translation*, by Paul Ricoeur, vii–xx. Translated by Eileen Brennan. New York: Routledge.

———. 2023. Anacarnation: Recovering Embodied Life. In *Anacarnation and Returning to the Lived Body with Richard Kearney*, ed. Brian Treanor and James L. Taylor, 233–256. New York: Routledge.

Keetley, Dawn. 2016. Without Name: Nature's Power. *Horror Homeroom*. https://www.horrorhomeroom.com/without-name/. Accessed July 20, 2024.

Onishi, Brian Hisao. 2020a. The Uncanny Wonder of Being Edible to Ticks. *Environmental Philosophy* 17 (2): 199–219.

———. 2020b. Weird Ecologies and the Uncanny in *The Happening*. In *Philosophy, Film, and the Dark Side of Interdependence*, ed. Jonathan Beever, 157–173. Lanham, MD: Lexington Books.

———. 2022. Weird Environmental Ethics: The Virtue of Wonder and the Rise of Eco-Anxiety. *SATS: Northern European Journal of Philosophy* 23 (1): 33–53.

Ricoeur, Paul. 2006. *On Translation*. Translated by Eileen Brennan. New York: Routledge.

———. 2007. *Hermeneutics and the Human Sciences*. Edited and translated by John B. Thompson. New York: Cambridge University Press.

Treanor, Brian. 2023. Thinking Like a Jaguar: Carnal Hermeneutics, Touch, and the Limits of Language. In *Anacarnation and Returning to the Lived Body with Richard Kearney*, ed. Brian Treanor and James L. Taylor, 13–31. New York: Routledge.

Without Name. 2016. Directed by Finnegan, Lorcan. Dublin, Ireland: Lovely Productions https://www.amazon.com/gp/video/detail/B0CVFZYBCS/ref=atv_hm_vid_c_baeECJ_1_5.

CHAPTER 7

(Eco)-Weirding Folk Horror in Alex Garland's *Men*

Dawn Keetley

When Alex Garland's *Men* was released in 2022, reviewers almost without exception dubbed the film 'folk horror'. *The Village Voice*, however, claimed the film as an example of a different and 'emergent' subgenre of cinematic horror—the 'new weird', going on to describe such films as those that use 'the tropes and impulses of horror to do … something else, mysteriously, metaphorically, and often inexplicably' (Atkinson 2022). Garland's *Men* does indeed stray into the terrain of the 'weird'. As a result, its status as folk horror becomes unsettled. At first glance, the film pursues what has become by now a familiar folk horror plot: an urban (even 'foreign'—she is Irish) character travels to an isolated English rural village and meets hostile locals who hold 'skewed' beliefs (Scovell 2017, p. 18). This plot, however, coexists with a 'weird' plot. Michael Cisco describes the weird plot in ways that uncover its uncanny mirroring of the folk horror plot. It is built, he writes, around 'the solitary individual who, cut off from institutions and thrown onto their own resources exclusively, confronts an event that seems to belong to what lies beyond the outer limit of human

D. Keetley (✉)
Lehigh University, Bethlehem, PA, USA
e-mail: dek7@lehigh.edu

B. H. Onishi, N. M. Bell (eds.), *The Call of the Eco-Weird in Fiction, Films, and Games*,
https://doi.org/10.1007/978-3-031-77126-2_7

understanding' (2021, p. 39). This, too, happens to *Men*'s protagonist in an entangled weird plot that renders the parallel folk horror plot profoundly strange.

There are several disruptive events in *Men*'s narrative that evince the 'waywardness of the veer' by which Roger Luckhurst characterizes the weird (2017, p. 1050). The first occurs when protagonist Harper Marlowe (Jessie Buckley) enters the local church and discovers a large font. Although almost every English church has a font, this particular font is in the wrong place and is inscribed with two folkloric figures that almost never appear on a font and certainly never together: the green man and the sheela na gig. But the *really* weird moments in *Men* turn out to involve Harper's repeated encounter with a naked man (Rory Kinnear), whom she first sees on her walk into the forest—a man Garland explicitly intended as 'a version of [a] Green Man' (Jenni Miller 2022). As the film progresses, greenery starts to sprout from his body, and, as it does, this man—a living version of the folkloric figure found on the font—becomes an increasingly visible incarnation of the weird, embodying what Cisco describes as 'the deterritorialization of ordinary experience, insofar as the ordinary is fetishized as "reality"' (2021, p. 7). The 'green man' challenges both the 'reality' that folk horror frequently presumes and 'reality' as Harper (and potentially the viewer, lured into identifying with Harper) understands it—specifically, a 'reality' constituted by particular conceptions of the rural and of bifurcated power relations between the urban and the rural that, in *Men*, also get transposed onto gender.

These weird moments in *Men* serve less to instantiate the film as an exemplar of the 'weird' genre—if there is such a thing—than to 'weird' the film, the folk horror genre, and Harper's worldview. As Luckhurst has pointed out, the weird is, paradoxically, 'a genre that dissolves generic glue' (2017, p. 1042); it is 'an inflection or tone, a *mode* rather than a genre' (2017, p. 1045). Luckhurst cites Veronica Hollinger's 'Genre vs. Mode', in which she posits that while genre designates an entity, mode signals movement. Genre, she writes, is a 'particular kind of narrative complex—generally understood to be an archive of stories with particular themes, motifs, and figures'. One only needs to think here of the multiple iterations of the supposedly defining characteristics of folk horror. 'In contrast', Hollinger continues, 'mode implies not a kind but a method, a way of getting something done' (2014, p. 140).[1] Hollinger's language ('getting something done') is perhaps too purposeful, too instrumental, for the weird-as-mode, which seems more accurately described by Elaine Joy as those 'propensities and relations' of a given text 'that might pull against the system and open it to productive errancy (literally "rambling," "wandering"—moments of becoming-stray)' (2013, p. 29). The embodied 'green man' of *Men* does indeed pull the text into errancy, making it stray, embodying the 'weirding' impulses of the film and resisting any efforts to make it *be* (generic) folk horror.

Pursuing those moments in *Men* that strain away from the systemic completeness and coherence of the folk horror genre, I will 'read for the weird', as Joy puts it, remaining 'open to incoherence' (2013, p. 30) and attempting to uncover 'the chatter and noise, the movements and operations, the signals and transmissions, the appearances and disappearances of the *weird* worlds, and their figures' (2013, p. 1; emphasis in original). My reading will be attentive to the weird as it exists 'in breach', as it 'twists or veers away from familiar frames and binary distributions' (Luckhurst 2017, p. 1052). To be attentive to the weird in *Men* is, moreover, to be of necessity attentive to the 'eco-weird', for it is the nonhuman that predominantly constitutes the 'waywardness of the veer' in this film (Luckhurst 2017, p. 1050), leading away from dominant meanings that are not only overdetermined but thoroughly anthropocentric.

7.1 *Men* as Folk Horror

Virtually every reviewer of Garland's *Men* identified the film as folk horror, describing it as a knowable entity in a clearly demarcated tradition—although, from the beginning, it is telling (given my argument that *Men* 'weirds' folk horror) that most reviewers felt the need to add an adjective to modify the film's 'folk horror'. Peter Debruge in *Variety* calls *Men* a

[1] In an important early essay, Alastair Fowler makes the persuasive case that genres are solidified in a certain historical moment while modes break free, evolve, and wander (like what Joy [2013, p. 29] calls the 'wandering' of the weird). Fowler writes: 'The genre … eventually exhausts its evolutionary possibilities. But the equivalent mode, flexible, versatile, and susceptible to novel commixtures, may generate a compensating multitude of new generic forms. For the mode was abstracted from an existing concrete historical genre. The latter, closely linked to specific social forms, is apt to perish with them. But the mode corresponds to a somewhat more permanent poetic attitude or stance, independent of particular contingent embodiments of it' (1971, p. 214).

'folk-horror bizart-house offering'; Jenni Miller in *Inverse* describes it as a 'stripped-down folk horror fable'; and Margot Harrison dubs it 'allegorical folk horror'.[2] Certainly, *Men* does (at first at least) unfold what is by now a recognizable folk horror plot. The film's protagonist, Harper, rents a house in an isolated rural village after the death of her abusive husband, James. We see flashbacks, from her perspective, of the two of them in their London flat arguing: Harper tells James she wants a divorce; James refuses to countenance Harper's leaving him, and he threatens to kill himself and then punches Harper, causing her to demand that he leave; James then falls past their apartment window, looking directly at Harper as he falls. She will subsequently tell the vicar of the rural village to which she flees that she heard, later, that James burst into an apartment above theirs and was trying to climb down onto their balcony when he fell, in what may have been an accident but may also have been suicide. Harper comes to the country, then, to heal.[3] As in much folk horror, the film depicts her car journey from city to country, and there are extensive scenes of her exploring the local countryside, scenes in which the land becomes a palpable actor in her drama of grief and recovery. Also as in the typical folk horror plot, before too long, a string of increasingly strange and antagonistic locals appears in her life. *Men* thus exemplifies Adam Scovell's important formulation of the 'Folk Horror Chain': it centers an isolated community (the village of Cotson), skewed moral beliefs (every villager Harper meets exists on a spectrum that runs from off-putting to hostile), with a culminating and violent 'happening/summoning' (2017, pp. 17–18) (Fig. 7.1).

Second only to its typical plot (modern, urban outsiders encroaching on the isolated, homogenous rural community), folk horror is defined by the presence of the 'pagan'—of ancient beliefs, of 'old gods'. The 'pagan' element of *Men* inheres in the green man and the sheela na gig on the font

[2] Guy Lodge in *The Guardian* sees *Men* as more integrally a part of the folk horror canon, writing that the film is 'rooted in a sturdy, shivery tradition of British folk horror, where the myths and traditions of the land become their own kind of threat', and he curates a list of folk horror films to watch with *Men* on the occasion of its release on streaming services, including the 'unholy trinity'—*Witchfinder General* (1968), *The Blood on Satan's Claw* (1971), and *The Wicker Man* (1973)—as well as *Cry of the Banshee* (1970), *Quatermass and the Pit* (1967), *Kill List* (2011), *A Field in England* (2013), and *The Hallow* (2015).

[3] The move to the country to 'heal' resonates especially with *Robin Redbreast* (1970), in which a woman retreats to the country after the break-up of a long-term relationship, and with *Wake Wood* (2009), in which a couple move to the country after the death of their daughter.

Fig. 7.1 The green man on the font in the village church (long shot and close-up)

in the village church. Indeed, Garland himself has claimed in numerous interviews that the green man and its folklore, its 'paganism', is the heart of the film for him.[4] 'I've been turning round that iconography for a long, long time', he said in an interview for *Vox*, continuing that its roots 'seem as likely to be pagan as anything else, and the fact is, nobody knows. Yet they clearly meant something important because they were carved into stone, over and over' (Wilkinson 2022). In installing the 'pagan' icons of the green man and the sheela na gig at the center of his film, Garland expressly engages, moreover, in the ***invention*** of folklore which is at the heart of folk horror. Folk horror may depict what appear to be 'authentic' traditions, rituals, and artifacts, but they are typically highly mediated and often expressly fabricated. Michael Dylan Foster and Jeffrey Tolbert have dubbed such fabricated folklore the 'folkloresque'—material inspired by folklore, 'imbued with a sense of "authenticity"', but invented nonetheless (Tolbert 2023, p. 34). As Scovell succinctly puts it, folk horror often 'creates its own folklore' (2017, p. 7).

[4] Rory Kinnear said in an interview, 'The Green Man was the one character that Alex [Garland] was absolutely driven by a certain vision. He was incredibly exacting about its look. He let me have the opportunity to flesh it out, to embody it—how he walked, how he behaved—but in terms of the look, it meant something incredibly powerful to him to the extent that for the last sort of stage of the makeup, he would do the makeup.

He had seen it so vividly over however many years that he's been thinking about Green Men for about 15 years he has been trying to write about it in some way. That was quite something to sort of see the level of care and concern he took over this figure' (Miller 2022).

In his very claim that the green man is 'pagan', Garland engages in the 'folkloresque' since the consensus in contemporary scholarship is that the green man is not in fact a 'pagan' survival in an evolving Christian modernity. An influential 1939 article by Lady Raglan claimed that 'unofficial paganism subsisted side by side with the official religion' (1939, p. 56)[5] and that the green man was part of a May-day 'cult' connected to such figures as Jack-in-the-Green, Robin Hood, and the King of May, and, through them, to 'the magical rite of the spring sacrifice' (1939, pp. 50, 54). Raglan's essay articulated what became the dominant view that, as Richard Hayman puts it, the green man signified the survival of 'pagan nature worship' in Christian culture (2010, p. 5). This claim has been debunked, however, as numerous subsequent historians have argued that the green man did not represent the persistence of paganism within an opposite Christianity but was instead mobilized by Christianity itself.[6] The generic impulse of folk horror, however, demands a central persistence of the 'old ways', which it articulates through its invented traditions—and *Men* deploys the figure of the green man to represent this survival of paganism, along with the 'sacrifice' which Raglan argues the green man inevitably evokes.

Robin Hardy's *The Wicker Man* (1973), a clear influence on *Men*, famously contributed to shaping the invented meanings of the green man as 'pagan' survival within folk horror, as the sternly Christian Sergeant Neil Howie (Edward Woodward) stays at the Green Man Inn upon arriving at Summerisle. He enters the pub just after a close-up of the sign swinging in the breeze; the head's vegetation is reproduced in the next long shot of the pub itself, which has greenery creeping over it. The virginal Howie will soon be assailed by a wild and unfettered sexuality both

[5] Raglan's larger argument has been fundamental to folk horror, which explains its recurrence in the genre: Dr. C. B. Lewis claims, Raglan begins, 'that "it would be a mistake to think that because Christianity finally triumphed in the long struggle with its pagan rivals, the latter disappeared the moment the former was officially recognized." The fact is that unofficial paganism subsisted side by side with the official religion, and this explains the presence of the Green Man in a church window with the Virgin beside him and below him the sun' (Raglan 1939, pp. 54–6). It is telling in light of Raglan's argument about the presence of paganism in Christianity that Garland expressly sutures together shots of the green man on the font with shots of Christ in the stained glass window of *Men*'s village church.

[6] For discussions of the green man, see Basford (1978), Coulter (2006), Hayman (2010), and 'Introducing the Green Man', https://blogs.loc.gov/folklife/2021/01/introducing-the-green-man/. I discuss the green man and its role in horror—specifically plant horror—in 'Introduction: Six Theses on Plant Horror' (Keetley 2016, pp. 1–5, 23–5).

Fig. 7.2 The Green Man Inn on Summerisle in *The Wicker Man*

inside and outside The Green Man: he will be subject to the temptations of the innkeeper's daughter (Britt Ekland) within, and he will witness multiple couples having sex just outside. This avatar of modern, urban Scotland (Howie flies over from the mainland), and of a strict orthodox Christianity, will see everything he stands for challenged by the opposed and expressly 'pagan' beliefs and practices of the inhabitants of Summerisle, all signaled by the figure of the green man; indeed, as friendly as the islanders seem, their intent in luring him to the island is to sacrifice him to a god rooted in the same natural world as the green man itself (Fig. 7.2).

In folk horror and its invented folklore (including *The Wicker Man*), the green man is associated not only with the survival of paganism, nature, and sacrifice but also with *reproduction*, which is another defining characteristic of folk horror. *Men* certainly represents reproduction and does so, like *The Wicker Man*, through its figurations of the green man. When Harper is out walking in the forest near her cottage, she encounters a naked man—one who will soon begin to sprout greenery and literally embody the enfoliated green man on the font. This encounter directly evokes the 1970 folk horror film, *Robin Redbreast*, in which Norah Palmer, recently removed from London to the country, encounters a virtually naked man (the Robin of the title) in the woods; he will later, as part of the ritual engineered by the locals, pursue and impregnate Norah. The naked man Harper sees will similarly follow Harper out of the woods, pursue her, and, late in the film, in a dramatic transformation of Norah's pregnancy in *Robin Redbreast*, will birth all the men that have harassed Harper, manifesting not the reproduction of children but of masculine power and violence.

7.2 *Men*'s Politics and Folk Horror's Anthropocentrism

The *seemingly* spectacularly 'weird' scene near the end of *Men*—the green man's birthing of all the village men (along with Harper's husband) who have harassed, stalked, and abused Harper—is made less weird because, like much of the film, the scene is, I argue, Harper's hallucinatory fantasy; it represents her trauma, her grief, and her efforts at healing. *Men* is, in the end, driven by Harper, by her particular perspective and by an anthropocentric view more broadly. The cinematography, for instance, relentlessly emphasizes Harper's perspective. When she is walking through the woods, the land around her is filmed in soft focus, with herself in the foreground, the only object in sharp focus, and almost always in the center of the frame.[7] After she reaches a long (disused) railway tunnel, Harper revels in making sounds and hearing them echo back to her. One commentator has argued that this is a moment when Harper 'can delight in the sound of her own voice, a voice which has been (and will continue to be) routinely belittled and ignored' (Acolytes 2022). While this is, on one level, certainly true, on another level, this scene literalizes the 'echo chamber' Harper inhabits, as she finds only what she expects in the countryside. *Men* thus shapes a particular psychological and political folk horror formation that is in conflict with the film's weirding tendencies. The fact that almost everything is shaped by Harper's gaze resists the 'weirding' work of the film, since the weird, as Joy aptly puts it, serves to 'unground texts from their conventional, human-centered contexts' and involves reading for things in texts that don't always 'humanistically re-boot the narrative' by referring back to such 'human-centered' constructs as 'human psychology' (2013, pp. 30–1). *Men*'s weird is challenged at every turn, then, by Harper's gaze; Harper's psyche is at the center of the film (Fig. 7.3).

Many of the film's folk horror motifs, moreover, are repurposed to a specific political end that is bound up with Harper's focalizing perspective, adumbrating the myriad of ways in which Harper is dismissed, belittled, harassed, and even assaulted by men at every turn. *Men* is a pitched battle between women and, well, men. Garland has said, 'I try to make films that have arguments, if you want them, but also don't have arguments if you don't' (Wilkinson 2022), which hints at the warring propensities in his

[7] The video essay, 'Men: Rhyme Over Reason' by Acolytes of Horror, does a great job of illuminating how Harper is always at the center of the shot in her walk through the woods—until the scene in which she reaches the tunnel and sees the man at the far end.

Fig. 7.3 Harper at the center of the shot; Harper runs upon a blocked-up, disused railway tunnel

film. Harrison adeptly articulates the film's 'arguments': 'On one side is a woman's nightmare of being assailed by an endless stream of men whose love is impossible to distinguish from hate and violence. On the other is an equally hyperbolic male nightmare of being discarded as redundant by women who no longer depend on men for survival, or perhaps even for reproduction' (2022). And Guy Lodge, in *The Guardian*, writes that the film 'probe[s] male toxicity and female endangerment with a distinctly #MeToo-era lens' (2022). Hewing to its folk horror impulses, *Men* repurposes the binaristic conflict at the heart of folk horror (between the urban outsider and the rural community), displacing it onto an equally rigid *gendered* structure: it's Harper versus an entire village of men, who turn out, in the end, to be the same man, offering up a sustained critique of patriarchy—specifically of men's violence toward women—and its continual reproduction. Indeed, *all men* here are the same—and not just the denizens of the rural village, on some other side of the line from the urban refugee. Harper's former husband—Black and, like her, a Londoner—becomes bound up with the rural white men, as the urban/rural structure is displaced by women/men.

Indeed, *Men* purveys a pure crystallized, *essential* form of what Scovell calls the isolated community of folk horror, which must be 'cut off from some established social progress of the diegetic world', an isolation that creates 'skewed belief systems and morality' (2017, p. 18). Elsewhere, I have called this the 'monstrous tribe' of folk horror, which denotes an entity that is more than a mere 'community'; rather, it signals a group of people 'bound together by shared (folkloristic) beliefs, traditions, and practices—a community bound so tightly, in fact, that it constitutes a "tribe"' (Keetley 2020, p. 11).[8] *Men* brilliantly comments on the very notion of the 'tribe' and the ways it is deployed in folk horror, first, by the choice to use one actor to portray all the men Harper meets in the isolated village of Cotson, and, second, by representing, in its eminently bizarre penultimate scene, all those men being born, one from the other, in a spectacular scene of like reproducing like—the inexorable reproduction of the same. In a critique of the ways in which folk horror depicts its 'community', Jamie Chambers has argued that what he calls the 'monstrous communal' of folk horror is an 'indistinguishable chorality' of the folk, going on to claim that 'the *perceived homogeneity* of supposedly bounded communities is positioned as sinister, uncanny, and unnatural', and that folk horror films 'frequently feature communities acting with *eerie singularity* of mind, flawlessly enacting complex conspiracies to outwit the individualist agency of metropolitan protagonists' (2022, p. 22; emphasis added). *Men* posits that the 'folk' are not just an 'indistinguishable chorality', not just homogenous, not just an 'eerie singularity' but *literally the same*, highlighting and undoing folk horror's representations of stereotypical, homogenous 'folk'.

Men's exploitation of the binary of folk horror in service of its '#MeToo-era' politics reaches its apotheosis when the naked enfoliated man (who has sprouted increasing amounts of greenery in each scene) gives birth to one of the village men Harper has encountered who, in turn, births another, who births another until, finally, Harper's dead husband emerges. This scene seems to encapsulate all the inexplicability and strangeness of the weird, apparently shattering its realism.[9] I would argue, though, that it instead marks a shift to a different register of *allegorical* realism. It is at

[8] My purposefully anachronistic use of a word steeped in settler-colonial relationships draws on Pippa Norris and Ronald Inglehart's definition of a 'tribe' as an 'imaginary community demarcated by signifiers of us versus them—our people versus the others'. It coheres around characteristics of social identity such as 'economic, religious, or blood ties, with a common culture and dialect' (2019, p. 7).

[9] H. P. Lovecraft has written that one of two principal directions that 'Serious weird stories' can take is to be 'made realistically intense by close consistency and perfect fidelity to Nature except in the one supernatural direction which the author allows himself' (2000, p. 61). In this scene, the potentially 'supernatural' intrudes, except there is little that is 'supernatural' (ghostly, incorporeal) about it, since the men pursuing Harper seem to take on the physical, embodied traits of her dead husband, his fatal injuries.

the same time the apotheosis of the film's overt political message and of Harper's psychological trauma: all men are the same, invested in reproducing masculine power, determined to bend women to their needs. As Claire M. Class notes (summing up the conclusions of most commentators), the naked man 'becomes an amalgam of all the men with whom [Harper] has had unpleasant run-ins'; as such, 'the figure takes on mythical interpretive proportions' (2022, p. 88). As the final man, Harper's dead husband James comes and sits on the couch with a resigned Harper who asks him (and, through him, all men), 'What is it that you want from me?' James replies, in the penultimate words of the film: 'Your love'. For Harper, then, the horror seems to be, as Wilkinson puts it, about 'whether there's any hope of escaping the insatiable neediness of men that thwarts the desires of women' (2022). This scene of reproduction, as ostensibly and graphically weird as it seems, is thus harnessed to a fairly straightforward and enduring allegorical message (as the vicar's quoting from William Butler Yeats' 'Leda and the Swan' and Samuel Daniel's 'Ulysses and the Siren' evinces):[10] Harper (and all women through time) continually have to confront men who project their own needs and desires onto them, who need, increasingly aggressively, to bend them to their will, to *make* them love them.

Men's birthing scene literalizes the previously strange sameness of the men Harper meets, their weird hostility, and their propensity (as in the case of the vicar, especially) to defend her husband, to suggest that *she* (not he) did something wrong, that she provoked his assault on her. This scene, in its literal reproduction of the strange sameness of men—suggesting that every man whom Harper has met is in fact in a conspiracy to entrap her—intersects with folk horror's tendency, to repeat Chambers' words, to depict communities that are an 'indistinguishable chorality', evincing 'an eerie singularity of mind' (2022, p. 22). The titular men are indeed, in the end, 'indistinguishable', of one mind, reproduced from one body. And the 'fertility' associated with the pagan in folk horror is thus repurposed to express the reproductive power of patriarchy and toxic masculinity, just as folk horror's sacrifice—the culminating 'happening/summoning' (Scovell

[10] The vicar quotes from 'Leda and the Swan': 'A shudder in the loins engenders there / The broken wall, the burning roof and tower / And Agamemnon dead'. And from 'Ulysses and the Siren': 'For beauty hath created been / To undo, or be undone'. He quotes both to convey to Harper that she is 'an expert in carnality' and that her sexuality (all imagined by himself) 'now exist[s] in my mind. This is the control that you exert'.

2017, p. 18) of folk horror—becomes the masculine demand that women sacrifice themselves, that they 'love' unconditionally and selflessly.

This particular political message of *Men* relies on reading the film through Harper's perspective; it relies on the interpretation that it is *through Harper's perspective* (not 'objective' reality) that all the men are the same. In the *LA Times*, Garland asked, 'Does Harper see all men as the same even though they're not? Or does she not realize that, in fact, they are all the same?' Garland continues, 'Those are wildly differently [sic] interpretations, and the question is in no way answered by the film' (Rottenberg). One scene in particular makes it clear, in my view, that it is *Harper* who sees all men as the same, that the men's culminating assault of her is a manifestation of *her* grief, trauma, and fear. In this scene, there are parallel shots of Harper on the floor in the corner of the kitchen—first, in a flashback, as she remembers being knocked by James's blow into the corner of the kitchen in their London flat, and second, as the reiterated attacks of the village 'men' begin. This parallel suggests that all the aggressiveness of the men represents echoes of James, traumatic after-shocks. This interpretation crystallizes as Harper asks James, 'What is it that you want from me?', and he says to her: 'My arm was ripped through by an iron railing. My ankle snapped. My internal organs crushed. This is what you did'. These things are precisely what Harper has (literally) also done to the 'men' in the village. She ripped open the hand and arm of one of them with a knife when he stuck it through her letterbox; and she hit another with her car, snapping his ankle and presumably crushing his internal organs. And what she did to one, she did to them all. Looking at his sliced-open arm and hand, the village boy says to Harper, 'You really hurt me. Look what you did'. These men are all, in the end, each other *and* James. Harper enacts what happened to James—what *he says* she did *to him*—on *their* bodies. In doing so, she learns to accept what happened to her husband and the fact that it was, in the end, self-defense. The scene ends with Harper holding the axe in her lap on the couch with the accusatory, needy, and abusive James next to her (Fig. 7.4).

Men definitely seems to have an argument, then, as Garland says his films can. The men in the village—the 'chorality' of hostile villagers (Chambers 2022, p. 22)—are all stand-ins for James, bodies on which Harper works through her guilt, her grief, and her rage. Caught in Harper's perspective, the green man—who ultimately generates them all—becomes a figure of the continual reproduction of men, specifically of masculine neediness, dependency, power, and violence. In the review that

Fig. 7.4 Harper at home after James hits her; Harper in her rental as she fears someone intrudes

suggests *Men* hints at a 'new weird', Atkinson goes on to argue that the film actually fails to live up to this suggestion, becoming instead 'a metaphoric ordeal, a masked procedural of trauma and healing'. He concludes that for 'all of its New Weirdness, the film proves to be ultimately conventional', and that 'Garland chooses the easy, fashionable route instead, and makes heaven and earth move for the sake of one woman's remorse' (2022). Reiterating an earlier point that 'the message' of 'the threat of men toward women' is 'rather heavy-handedly made', Diane Rodgers concurs, asking, in her review of the film, 'Is the metaphor heavy-handed? Yes. Does the folk-horror narrative do something new? Not especially' (2024). *Men* does do something new, though, in moments when the film escapes Harper's psyche, when it 'weirds' its otherwise formulaic folk horror and its conventional and 'heavy-handed' gender politics. As Brian Hauser has pointed out, the weird resists 'reduction to allegory', must, in fact, specifically avoid 'allegorical displacement' (2018, pp. 240, 243). And *Men* does offer this resistance—specifically in those moments when it veers toward the ecological weird, moments which are more numerous than Atkinson credits.

7.3 Eco-Weirding Green *Men*

While the carved green man and sheela na gig on the font in the village church serve as ostensible pagan 'survivals' that are assimilated into the overdetermined 'folkloresque' of folk horror and the 'heavy-handed' allegories of the reproduction of toxic masculinity, what actually persists as 'weird' is the naked man himself—the embodiment of a green man—whom Harper encounters when out walking in the forest. This weird 'veer' of *Men* is signaled by the green man's very visible imbrication with nature (vegetation literally grows from his body as the film progresses), by his obvious class difference, and by the location in which he is found and with which he remains associated (and in which parts of *Men* were filmed).

He is first discovered by Harper alone in the woods. As she emerges from the woods, Harper turns to take a picture and he comes into view, next to a tree and some ruined buildings. As he, the tree, and the buildings come into focus, Harper goes out of focus, dramatically reversing the unrelenting focus on her (and the out-of-focus natural surroundings) as she had walked in the woods in the prior scene. The next shot is an even longer shot of the man, who thus becomes even more enmeshed with the trees and ruined buildings. Immediately, the human—and, specifically, Harper's totalizing gaze—is dislodged (Fig. 7.5).

Harper hurries away, disconcerted, and her next encounter with the man is when she sees him in the garden of the house she's renting, staring at the apple trees; his subsequent appearance (this time not from Harper's point of view) is in what seems a kind of outdoor shrine, where he inserts leaves into cuts on his skin. Later, the green man appears again in front of Harper and blows dandelion seeds at her; finally, near the end, he appears as a full-fledged enfoliated green man at the beginning of the birthing scene. In all of these scenes, he embodies what Timothy Morton calls the constantly changing 'mesh' of 'lifeforms and non-life', an ecological system of interconnected objects, of 'unique irreducible strange strangers' that are 'both human and nonhuman' (2011, p. 185). Luckhurst has claimed the tentacle as the 'signature of weird fiction and horror film', its 'limb-tongue suggestive of absolute alterity' (2017, p. 1054), but tendrils, shoots, and vines—especially when they are in unexpected places doing unexpected things—can also represent such 'absolute alterity'.[11] In *Men*, they serve to distance the film still further from Harper's perspective, from the folk horror plot, from the political allegory of the reproduction of men's toxic power over women, and from anthropocentrism more generally. They signal the weird's 'deterritorialization of ordinary experience' (Cisco 2021, p. 7) and, more specifically, what Brian Onishi has defined as

[11] See Keetley, 'Introduction: Six Theses', pp. 6–9, and Scherer, 'Pre-Cosmic Squiggle', 2016, pp. 31–53.

Fig. 7.5 Harper's first encounter with the 'real' naked green man

the eco-weird—that which 'designates nature as uncanny irruption' (2022, p. 169). Lawrence Buell has asserted that the effect of environmental consciousness is 'to raise the question of the validity of the self as the primary focalizing device for both writer and reader'—shifting the perspective to something less human (1996, p. 179). The green man—the 'real' green man—is the figure that effects this shift in *Men*, moving the focus from Harper and Harper's worldview (and, indeed, from the self as the 'primary focalizing device' at all) to something else, something nonhuman (Fig. 7.6).

A shot of the green man in his shrine, inserting leaves into his skin, is part of an important montage at around halfway through *Men*, one that markedly detaches the camera from Harper's point of view (or any other character's point of view, for that matter). This montage is composed of objective shots of the land—of fields, woods, and sky; it includes a long shot of a dead deer, tracking (along with a dandelion seed) into its eye and then out after time has clearly lapsed, as maggots now move over the carcass. There is no evident metaphorical meaning for these shots (a comparison to the overdetermined deer imagery in Jordan Peele's 2017 film, *Get Out*, is illustrative here, for instance). Instead, they serve only to detach the camera from Harper's perspective and to center the nonhuman in the forest—dying, decomposing, consuming, living (Fig. 7.7).

Fig. 7.6 The green man in a kind of shrine and, later, as fully enfoliated

Fig. 7.7 The deer before the tracking in and after the tracking out shots

As well as being linked with a nature that escapes the human perspective, the green man represents a different way of living in nature than the one Harper's perspective crafts for much of the film. This way of living constitutes another way in which *Men* 'weirds' the expected folk horror narrative, as the film becomes not a mimetic rehearsal of notions of 'landscape' and the 'rural' that are typically purveyed by folk horror but an interrogation of them, challenging its regressive nostalgia and its suppression of real places in service of a generic and idealized 'landscape'. Unearthing these meanings depends on a reading that attends to the actual landscape, including the filming location, in which the living green man appears—and that is, the Forest of Dean. The green man's appearance not just in woods, moreover, but in a ruined structure on the edge of woods evokes a long history of living and working in the forest that explicitly counters the creation of the 'Cotswold village' as tourist site (the other primary filming location); together, these two locations offer two vastly different ways of organizing work and cohabiting with nature. The 'weird' in *Men* not only challenges the 'nature of reality' (Cisco 2021, p. 27), then, but a specifically constructed form of reality—the 'Cotswold village', a construction that exists at the expense of another and 'wilder' way of living, one long manifest in the Forest of Dean.

Although landscape is routinely posited as a central characteristic of folk horror, it is frequently an abstraction, a critical concept imposed on the actual setting and location of the film. As such, it paradoxically obscures the *place* of the film even as 'landscape' seems to be centered. And when (real) place is dis-placed, so are the people who live and work on it. As cultural geographer James Thurgill writes, the eeriness of folk horror's landscapes derives from 'a deliberate attempt to exploit the *othering* process manifest in the presentation of rural communities as something outside of the normative'. Typically observed through the 'urban gaze' (as we see in *Men*'s persistent privileging of Harper and her perspective), the 'nuance of pastoral landscape is lost in reductive articulations of rurality as singular and static' (Thurgill 2020, p. 34).[12] While *The Blood on Satan's Claw* famously begins, for instance, with a scene of a man plowing a field, his work soon stops (and stays stopped) as he unearths a demonic skull, problematizing (to say the least) Paul Cowdell's claim that 'Folk horror films are predicated on more complete visions of socio-economic worlds within a time and space. Farming is an actual practice in folk horror, rather than a pretextual backdrop' (2019, p. 301). Folk horror, rather, tends to offer a 'landscape' that is often portentous with the demonic, shaping an 'aesthetic experience of exoticism' that is not 'picturesque' but 'horrifying, threatening, and abject' (Chambers 2022, p. 26). Indeed, as Peter Hutchings notes, landscapes that 'form part of a dark heritage' are 'in

[12] Chambers also argues that folk horror inherently erases the specificities of space and rural communities, largely through the process of 'exoticism and cultural appropriation' (2022, p. 11).

their own way just as picturesque and marketable as their more conventionally pretty counterparts' (2004, p. 28).

Men exploits and purveys this doubled notion of the picturesque; it is set in a 'perfect' village and filmed in Withington, Gloucestershire, itself an 'idyllic' Cotswold village, yet it turns into the menacing locale of alien, othered and finally violent villagers. Harper notably flees from London to a 'rural' that at first conforms exactly to what she expects. She finds the ideal country house embedded in the beautiful, lush, isolated countryside, where we see moments of joy that indicate the 'healing' she went to the country to experience. That Harper finds only her own prior expectations about the pastoral, however, is amplified in a scene in which she gives her friend Riley a 'tour' of the house via her phone, which becomes ironic, even parodic, because it is exactly what they both expect of the country house, with its Aga range, its Shaker walls, and its 'baby grand'. Mark Kermode observes in *The Guardian* that Harper 'has escaped into verdant surroundings for a fortnight in "the dream country house"—with the emphasis on "dream"'. And he notes 'the oversaturated palette of Rob Hardy's cinematography (fields so green they glow, flowers that pop in purple-blues)' and 'the bloody, chocolate-box interiors conjured by production designer Mark Digby and set decorator Michelle Day' (2022). If this house, this countryside, is a dream, it's Harper's dream, but it also limns the fantasy of all those who flee London to 'the country'—specifically, in this case, to the highly idealized rural villages of the Cotswolds.

The two primary Gloucestershire filming locations, Withington and the Forest of Dean, represent the stark contrast between the idealized place and the place that evades the ideal. Historian Ralph Anstis notes that the Forest of Dean 'is at the unfashionable end of Gloucestershire, overshadowed by the better-known Cotswolds at the eastern end of the county' (1986, p. 3). And Michael Phelps writes that the Forest of Dean 'is a working forest and the site of many industries which have left their scars; some are unavoidable and many have been successfully healed by nature. Nevertheless, cosseting, cherishing and preserving can be carried to extremes, only choking the thing which is loved. There are beautiful Cotswold villages, perfectly kept, all they lack is life; whatever anyone may say about Forest villages, that cannot be said' (2008, loc 216–19). It is significant to the 'weird' impulses of *Men* that the naked green man originates outside of the idealized village and away from the 'perfect' country house. Indeed, he is first seen in the forest and in close proximity to a ruined reminder of the working history of the Forest of Dean.

Indeed, this very different 'rural' intrudes into the film even earlier than Harper's sighting of the naked man, when she is on her walk in the forest, just prior to that meeting, and she finds a disused railway tunnel. And then, as she leaves the tunnel and heads home, unnerved by having seen a man at its other end, she encounters other ruined industrial buildings on the edge of the forest. Both the railway tunnel and the abandoned buildings invoke the long history of industry in the Forest of Dean. There has been mining (primarily of iron ore and coal) in the forest since the Celts (Phelps 2008, loc. 104) or Silures (Anstis 1986, p. 4) lived there. Indeed, the Forest boasts a long tradition of free mining, as inhabitants of and workers in the forest have for centuries declared their right to govern and work for themselves, resisting the encroachments of outside government and outside interests intent on exploiting their labor and the forest's natural resources.[13] Disused tramroads, railways, and tunnels dot the forest, overgrown by nature, as evidence of the heyday of coal mining in the eighteenth and nineteenth centuries—and its subsequent demise (Phelps 2008, loc 244). If one reads *Men* through a 'historically tethered, particularist approach' (Chambers 2022, p. 28), its filming locations offer glimpses of a forest and a rural existence that also function as a weird 'veer' away from the ideological construction of the (singular) rural that typically characterizes folk horror.

Of critical importance in recognizing a buried and place-based history of class and work in *Men* is the scene in which Harper first encounters the naked green man by a ruined building on the edge of the forest. Right before Harper sees the man, she walks past the ruined buildings he stands near—buildings that seem to confuse and unnerve as she looks and then hastens past. Indeed, the buildings disturb both the idealized village and pastoral woods settings she has been enjoying thus far. As the camera lingers on the exterior of these buildings, and even takes up a position inside briefly, we get a glimpse of an industrial history, of labor that was present but has since moved on. Harper is subordinated in this brief scene, especially when the camera sets up within the building and she becomes relegated to a non-individuated figure on the edge of the frame. That the naked man then appears in front of these buildings—each of them (man *and* building) disrupting the photograph of the forest Harper is trying to

[13] See Anstis (1986) for the history of the Dean Forest Riots of 1831, led by hero/rebel Warren James. See Hart (2005) for a history of free mining in the forest. And see Chris Fisher (1981) for a Marxist history of the Forest of Dean colliers between 1788 and 1888.

Fig. 7.8 Harper hastens past the ruined buildings

take—invokes a rural working-class reality that disrupts everything Harper has so far encountered in her countryside retreat. After the naked man appears in Harper's garden, she immediately calls the police who presume he's been 'sleeping rough' in the woods. The police officer later suggests he's a 'gyppo'. These efforts at dismissal by the inhabitants of the idyllic 'imagined village',[14] however, can't dispel the Forest of Dean's history of working people introduced by Harper's encounter with the green man (Fig. 7.8).

That history is ushered in not only by the man himself but by the ruins with which he is associated—ruins which certainly serve to 'weird' Harper's rural ideal. The 'ruins' of *Men*—the abandoned railway tunnel and the ruined (industrial) buildings—can be situated in a weird tradition that has long centered ruins and powerful nonhuman objects, and that has, in doing so, also destabilized the human-centered folk horror narrative. I would argue that the strange ruins scattered through the forest in *Men*, with their industrial history, are akin to the ruins that feature in weird fiction such as Arthur Machen's 'The Shining Pyramid' (1895), E. F. Benson's 'The Temple' (1924) and 'The Flint Knife' (1929), Robert E. Howard's 'The Black Stone' (1931), Stephen King's 'N' (2008), and Adam Nevill's

[14] Georgina Boyes (1993) has described the 'imagined village'—a concept integral to many folk horror narratives.

Fig. 7.9 The weird ruins in the Forest of Dean: two blocked-up, disused railways tunnels, like that which features in Men; tram tracks swallowed by dirt and leaves; an abandoned smokestack (pictures are my own)

The Ritual (2011). Indeed, Phelps has written of the ruins of the Forest of Dean in ways that resonate with such tales of megaliths and stone altars that are integral to the weird literary tradition: 'Past and present are interwoven here—that which was thought to be past and gone is suddenly found to be present and alive, one is not even always sure which tense to use when talking about the Forest' (Phelps 2008, loc. 72). Phelps adds: 'If we came upon a group of Druids, Silures or Romans it would not be such a surprise here as it would anywhere else' (2008, loc. 73). The ruins of the Forest of Dean, standing alone in nature, wrenched from—abandoned by—their working-class history, are certainly as alien—as weird—to Harper and her expectations of 'dream country homes' and Aga ranges as a Druid stone circle would be (Fig. 7.9).

Mark Fisher makes an important point about the weird when he argues (quoting Maurice Lévy) that it is 'a fiction in which, not the impossible but the *outside* "can make an irruption, through time and space, into an objectively familiar locale"'. It is, he continues, 'the irruption into *this* world of something from outside which is the marker of the weird' (2016, p. 20). *Men* suggests that this irruption from the outside, that which 'deterritorializes' ordinary experience 'insofar as the ordinary is fetishized as "reality"' (Cisco 2021, p. 7), need not 'irrupt' from some utterly inexplicable 'impossible' exterior realm—a narrative event that might be said to constitute the specifically cosmic weird rather than the merely weird.[15] Instead, 'weirdness' may irrupt from a version of a 'real' that has been thoroughly erased—suppressed to such an extent that it *becomes* 'weird'. Brian Onishi (2023, p. 133) has aptly described this way of conceiving the

[15] See Onishi (2022, p. 48) for a definition of the specifically *cosmic* weird, 'where the intrusion manifests from some other dimension or reality' which I'm distinguishing from what I am describing here—the merely weird.

eco-weird in particular, arguing that, in the eco-weird, 'it is not that which is outside our reality that breaks in, but it is what lies underneath our surface experience that emerges'. He calls this the eco-weird's impulse toward 'extrusion' (rather than intrusion)—'where reality is pushed to the surface through the veil of appearance'. This weird 'extrusion' can be of a now-strange past, of long-dead peoples, and of the remains of their living and working—all of which 'extrude' through the flat, idealized structuring constructs of both folk horror and certain versions of the real.

In an essay cited by Robert Macfarlane (2015) in his well-known 'The Eeriness of the English Countryside' (an article foundational within folk horror studies), commentator Nick Groom 'deplores the "pastoral clichés that have tyrannised the land for decades"; we need "to start rethinking what we understand by the English countryside," he urges, "and we need to be doing it now"'. Macfarlane argues that this 'political "rethinking"' is apparent in some of the 'eerie culture' that has so often—since 2015—been gathered under the label 'folk horror'. But as I have argued in this chapter, to *really* effect this 'rethinking' of the 'English countryside', and of anthropocentrism more generally, we need to 'weird' the folk horror tradition itself. Folk horror has tended to rely on those 'pastoral clichés' deplored by Groom—clichés driven by nostalgia, idealization, and those 'tendencies towards exoticism and cultural appropriation' that Jamie Chambers (2022, p. 11) identifies within so many of the genre's key texts. To 'weird' folk horror is to excavate all that is beyond its reified aesthetic and epistemological structures—that which 'extrudes' (Onishi 2023, p. 133) from outside, from underneath. To 'weird' folk horror is to recognize the dominant work of 'exoticizing' and 'othering' in which the genre frequently engages, but it is also to recognize disruptive impulses, including those that are directed at folk horror's anthropocentrism, its nostalgic idealization, and its romanticized rural dreamscapes.

Acknowledgments I would like to thank Lehigh University for a generous Faculty Research Grant that made this chapter possible.

References

Acolytes of Horror. 2022. Men: Rhyme Over Reason. *YouTube*. https://www.youtube.com/watch?v=VbJNgJuu_zY. Accessed 2 June 2024.

Anstis, Ralph. 1986. *Warren James and the Dean Forest Riots: The Disturbances of 1831*. Gloucester: Alan Sutton Publishing.

Atkinson, Michael. 2022. The New Weird: The Mystery of Alex Garland's 'Men'. *The Village Voice*. 24 May. https://www.villagevoice.com/2022/05/24/the-new-weird-the-mystery-of-alex-garlands-men/. Accessed 2 June 2024.

Basford, Kathleen. 1978. *The Green Man*. Suffolk UK: D. S. Brewer.

Boyes, Georgina. 1993. *The Imagined Village: Culture, Ideology and the English Folk Revival*. Manchester: Manchester University Press.

Buell, Lawrence. 1996. *The Environmental Imagination: Thoreau, Nature Writing, and the Formation of American Culture*. Cambridge, MA: Harvard University Press.

Chambers, Jamie. 2022. Troubling Folk Horror: Exoticism, Metonymy, and Solipsism in the 'Unholy Trinity' and Beyond. *JCMS: Journal of Cinema and Media Studies* 61: 9–34.

Cisco, Michael. 2021. *Weird Fiction: A Genre Study*. New York: Palgrave.

Class, Claire M. 2022. Scrutable Monstrousness in Alex Garland's *Men* (2022). *Studies in the Fantastic* 13: 85–90.

Coulter, James. 2006. *The Green Man Unmasked: A New Interpretation of an Ancient Riddle*. Bloomington, IN: AuthorHouse.

Cowdell, Paul. 2019. Practicing Witchcraft Myself during the Filming: Folk Horror, Folklore, and the Folkloresque. *Western Folklore* 78: 295–326.

Debruge, P. 2022. 'Men' Review: Alex Garland's Gender-Upender Fails to Earn Its Intense Body-Horror Finale. *Variety*. 9 May. https://variety.com/2022/film/reviews/men-review-alex-garland-1235261621/. Accessed 3 June 2024.

Fisher, Chris. 1981. *Custom, Work and Market Capitalism: The Forest of Dean colliers, 1788–1888*. London: Croom Helm.

Fisher, Mark. 2016. *The Weird and the Eerie*. London: Repeater Books.

Fowler, Alastair. 1971. The Life and Death of Literary Forms. *New Literary History* 2: 199–216.

Groom, Nick. 2013. 'Let's Discuss Over Country Supper Soon': Rural Realities and Rustic Representations. *The Clearing*. 22 August. https://www.littletoller.co.uk/the-clearing/lets-discuss-over-country-supper-soon-rebekah-brooks-and-david-cameron-rural-realities-and-rustic-representations-nick-groom/. Accessed 3 June 2024.

Harrison, Margot. 2022. The Allegorical Folk Horror of Alex Garland's 'Men' Will Fascinate Some and Frustrate Others. *Seven Days*. 25 May. https://www.sevendaysvt.com/vermont/the-allegorical-folk-horror-of-alex-garlands-men-will-fascinate-some-and-frustrate-others/Content?oid=35648344. Accessed 3 June 2024.

Hart, Cyril. 2005 (1953). *The Free Miners of the Royal Forest of Dean and Hundred of St. Briavels*. Lydney, Gloucestershire: Lightmoor Press

Hauser, Brian R. 2018. Weird Cinema and the Aesthetics of Dread. In *New Directions in Supernatural Horror*, ed. Sean Moreland, 235–252. Palgrave Macmillan.

Hayman, Richard. 2010. *The Green Man*. Oxford: Shire Publications.
Hollinger, Veronica. 2014. Genre vs. Mode. In *The Oxford Handbook of Science Fiction*, ed. Rob Latham, 139–151. New York: Oxford University Press.
Hutchings, Peter. 2004. Uncanny landscapes in British film and television. *Visual Culture in Britain* 5: 27–40.
Joy, Elaine A. 2013. Weird Reading. *Speculations: A Journal of Speculative Realism* IV: 28–34.
Keetley, Dawn. 2016. Introduction: Six Theses on Plant Horror; Or, Why Are Plants Horrifying? In *Plant Horror: Approaches to the Monstrous Vegetal in Fiction and Film*, ed. Dawn Keetley and Angela Tenga, 1–30. New York: Palgrave Macmillan.
———. 2020. Defining Folk Horror. *Revenant: Critical and Creative Studies of the Supernatural* 5: 1–32.
Kermode, Mark. 2022. Men Review—Alex Garland's Rural Retreat into Toxic Masculinity. *The Guardian*. 5 June. https://www.theguardian.com/film/2022/jun/05/men-review-alex-garland-jessie-buckley-rory-kinnear. Accessed 3 June 2024.
Lodge, Guy. 2022. Men and the Best British Folk Horror Films. *The Guardian*. 27 August. https://www.theguardian.com/film/2022/aug/27/streaming-men-and-the-best-british-folk-horror-films. Accessed 3 June 2024.
Lovecraft, H.P. 2000 (1927). *The Annotated Supernatural Horror in Literature*, ed. S.T. Joshi. New York: Hippocampus Press.
Luckhurst, Roger. 2017. The Weird: A Dis/orientation. *Textual Practice* 31: 1041–1061.
Macfarlane, Robert. 2015. The Eeriness of the English Countryside. *The Guardian*. 10 April. https://www.theguardian.com/books/2015/apr/10/eeriness-english-countryside-robert-macfarlane. Accessed 30 May 2024.
Miller, Jenni. 2022. The Stars of *Men* Unpack Its Mysterious Ancient Symbolism. *Inverse*. 19 May. https://www.inverse.com/entertainment/men-explained-sheela-na-gig-green-man-alex-garland. Accessed 2 June 2024.
Morton, Timothy. 2011. Waking Up Inside an Object: The Subject of Ecology. *English Language Notes* 49: 183–191.
Norris, Pippa, and Ronald Inglehart. 2019. *Cultural Backlash: Trump, Brexit, and Authoritarian Populism*. New York: Cambridge University Press.
Onishi, Brian. 2022. Weird Environmental Ethics: The Virtue of Wonder and the Rise of Eco-anxiety. *SATS* 23: 33–53.
Onishi, Brian Hisao. 2023. *Weird Wonder in Merleau-Ponty, Object-Oriented Ontology, and New Materialism*. New York: Palgrave Macmillan.
Phelps, Michael. 2008. *The Forest of Dean*. Stroud, Gloucestershire: Amberley, Kindle.
Raglan, Lady. 1939. The 'green man' in Church Architecture. *Folklore* 50: 45–57.

Rodgers, Diane. forthcoming, 2024. *Men*: A Folk Horror Gorefest from Writer-Director Alex Garland. *Revenant: Critical and Creative Studies of the Supernatural*, forthcoming

Rottenberg, Josh. 2022. Don't Expect Alex Garland to Explain What's Happening in 'Men'. But Here Are a Few Hints. *LA Times*. 20 May. https://www.latimes.com/entertainment-arts/movies/story/2022-05-20/men-explained-a24-horror-alex-garland-jessie-buckley. Accessed 3 June 2024.

Scherer, Agnes. 2016. The Pre-cosmic Squiggle: Tendril Excesses in Early Modern Art and Science Fiction Cinema. In *Plant Horror: Approaches to the Monstrous Vegetal in Fiction and Film*, ed. Dawn Keetley and Angela Tenga, 31–53. New York: Palgrave Macmillan.

Scovell, Adam. 2017. *Folk Horror: Hours Dreadful and Things Strange*. Leighton Buzzard: Auteur Publishing.

Thurgill, James. 2020. A Fear of the Folk: On *topophobia* and the Horror of Rural Landscapes. *Revenant: Critical and Creative Studies of the Supernatural* 5: 33–56.

Tolbert, Jeffrey A. 2023. The Frightening Folk: An Introduction to the Folkloresque in Horror. In *Folk Horror: New Global Pathways*, ed. Dawn Keetley and Ruth Heholt, 25–41. Cardiff: University of Wales Press.

Wilkinson, Alissa. 2022. From Ex Machina to Men, Alex Garland Is Always Exploring Original Sin. *Vox*. 17 May. https://www.vox.com/23086827/men-alex-garland-annihilation-eden-devs-machina. Accessed 4 June 2024.

CHAPTER 8

Staying with the Weird: Apophatic Wonder and Cosmographic Exploration in Eco-Weird Games

Nicholas J. Mizer

8.1 Introduction

In a partial fulfillment of Terence McKenna's prophecy (1998) that the world would finally get weird enough that the people would have to talk about it, the concept of the eco-weird emerged from the West Pennsylvanian darkness outside Brian Onishi's home in the ominous year of 2020 (Genovese 2023; Onishi 2020). From a noisome darkness of occulted plants "with their unseen networks of communication, their slow movement toward water, their vast time scales," tiny monstrous beings swarmed forth to incessantly tap against his quotidian screen door, illuminated by a domestic porch light that "reveal[ed] them in all their strange glory" (Onishi 2020, p. 157). Reflecting on the uncanniness of these bugs and plants and drawing on his experiences of Weird fiction, Onishi playfully imagined the former in terms of the latter. This was an act of apocalypse, in the original sense of unveiling: the eco-weird can "*reveal* the horrible

N. J. Mizer (✉)
Rensselaer Polytechnic Institute, Troy, NY, USA

B. H. Onishi, N. M. Bell (eds.), *The Call of the Eco-Weird in Fiction, Films, and Games*,
https://doi.org/10.1007/978-3-031-77126-2_8

wonder, and the terrible awe of being with others" (Onishi 2020, p. 172, emphasis mine). This was not the first time anyone noticed a connection between ecologies and horror fiction, whether we consider Donna Haraway's "Chthulucene" (2016), Smith and Hughes' "Eco-gothic" (2013), or the ways that the ecological violence of modernity shaped the emergence of the horror genre in the first place. But staying with the weird, rather than subsuming it to the Gothic or invoking it and disclaiming it for the chthonic, allows the eco-weird to offer unique insights as we continue to wend our twisting way through the apocalypse. The weird darkness of Onishi's eco-weird forest, permanently unilluminated outside the glow of the porch light, contrasts with the shadows and moonlight of the Gothic. The Gothic monster, once illumined by the light of rationality, retreats into the mind. The weird, no matter how much we expand our understanding, remains occluded, outside, beneath, and within all that we perceive and all that we cannot perceive.

Staying with the weird, following its twists and turns through ecological ethics (Onishi 2022) and ontology (Onishi 2023), has led to the particularly useful fruit of "Weird wonder," "the open and generative possibility that stands at the superposition of relation before collapse" (Onishi 2023, p. 121). Weird wonder is both a non-anthropocentric, ontological state of reality (the world *is* weirdly wondrous) and an operative virtue that can guide human imagining and acting in the world (our imaginations and actions in the world *should* be weirdly wondrous). Virtuous Weird wonder is balanced between the excess of (eco)anxiety and the deficiency of (eco)boredom. Both extremes paralyze our ability to respond to the often overwhelming challenges of climate change. This situates Weird wonder in the context of an environmental virtue ethic which "seeks the flourishing and good of both humans and nonhumans separately and together" (Onishi 2022, p. 34). A corollary of this framing is that we can develop judgments about conceptions and utilizations of Weird wonder based on their effectiveness at staving off both (eco)anxiety and (eco)boredom in ways that help us imagine and shape the world toward thriving multi-species fields.

In this chapter, I argue that Weird wonder is particularly effective when apophatic awe intertwines with unspeakable horror in the context of cosmographic play. To this end I begin by examining the connections between apophatic theology, Marxism, and the eco-weird, arguing for a conception of Weird fiction that expands outside the bounds of horror. From there, I sketch a model of tabletop role-playing games (TRPGs) as cosmographic technologies that assist weirdly wondrous imagination and praxis. I close by considering the games of David Blandy and his concept of Weirdhope as particularly effective case studies in the wonder-cultivating potential of eco-weird games.

8.2 Weird Apophasis

Onishi shudders at the plants slowly squirming in the darkness, but shuddering with horror can pass easily into trembling with awe and back again. The Latin *tremere* gives us both "tremble" and "tremendous," in the sense of to be trembled at," and the two meanings come together in Rudolf Ottos's *mysterium tremendum* (1923). Otto describes *mysterium tremendum as* an emotional response to absolutely unknowable alterity, a feeling that "may burst in sudden eruption up from the depths of the soul with spasms and convulsions, or lead to the strangest excitements, to intoxicated frenzy, to transport, to ecstasy. It has its wild and demonic forms and can sink to an almost grisly horror and shuddering" (Otto 1923, p. 12). In this passage, Otto's desire to describe the indescribable leads him toward purple prose that would be right at home in *Weird Tales.*

In gesturing toward this connection between the weird and the holy, I am deeply indebted to Taylor Genovese's synthesis of apophatic Marxism and the uncanniness of the weird into "a dialectical helix that spirals through [his] work" (Genovese 2023, p. 2). Both threads of Genovese's synthesis are influenced by China Miéville, although Miéville does not seem to have explicitly connected his influential definition of Weird fiction (2009) with his call for the development of apophatic Marxism (2019). In the latter work, Miéville argues for "a red *via negativa*, for which the unsayable entwines the apocalyptic veil-rending of the world with the utopian horizon glimpsed through the rip" (2019), resonating with Onishi's description of wonder as a virtue that "embraces the wounds of the world as open possibilities rather than debilitating obstacles" (2022, p. 18). Genovese's synthesis of the apophatic weird connects eco-weird wonder with a deep tradition of thinking and praxis devoted to the practical development of operative wonder.

In theology the apophatic, or *via negativa*, is contrasted with the cataphatic, or *via positiva*. Cataphatic theology proceeds by making positive statements about the nature of the divine, expanding human knowledge. Apophatic theology, by contrast, holds that any statement about God must be so limiting as to be false, that God is literally unthinkable. Instead, the *via negativa* embraces "the cloud of unknowing," in language almost indistinguishable from Onishi's description of "the comfort of an unknowing darkness in its secretive nurturing" (2023, p. 128).

Although figures like Pseudo-Dionysius, Nicholas of Cusa, and Thomas Merton have written influential Christian apophatic texts, it is important to note that the apophatic extends well beyond the bounds of particular religious traditions. *Apophasis* literally means "unsaying." As Michael Sells (1994) notes, if we define this unsaying in formal terms, it can be found in a wide range of sources, not the least of which is the *Tao Te Ching*'s "The *tao* that can be spoken is not the *tao*." Because I am writing as a descendant of French colonizers who were in turn descended from people colonized by Roman colonizers, however, I believe there is a wisdom in following Sellis, Miéville, and Genovese in confining my discussion to so-called Western sources. There, apophaticism first comes into focus with Plotinus, whose weird apophatic loops lead him to unsay even his unsaying, creating what Sells calls "an open, never ending process of apophasis" (1994, p. 17). Noting the infinite depths of that darkness and its richness for future unsaying of eco-weird Wonder, I stop just short of jumping into that abyss and instead offer the suggestion that the *via negativa* can help us shift our conceptions of the eco-weird away from a preoccupation with horror. In the words of Genovese

> this kind of aw(e)fulness is how Otto describes 'the holy' which he argues is too often attributed to moral perfection, when in actuality, it must be understood as a truly 'negative' experience—not only in the affective sense, but also as something profoundly ontologically apophatic. The holy is not natural to humans, *nor is it for us.* It is telling that, scripturally, angels first greet humanity with the words 'be not afraid'. (2023, p. 39)

Yet the approach toward the weird so often veers off into the merely horrible. The heart of Weird fiction, Miéville tells us, is "*awe*, and its undermining of the quotidian" (2009, p. 510). Awe, like many denizens of weird worlds, is a hybrid being, however, slithering between the awesome and the aw(e)ful. This is Kant's sublime, which he describes as "enjoyment but with horror" (1991, p. 47). However, conceptions of the weird tend to lean heavily on the horror and light on the enjoyment. Perhaps this is because of the preeminence of H.P. Lovecraft, whose "Supernatural *Horror* in Literature" (1945) forms the prototype of our understanding of the weird. For Lovecraft, the darkness looming outside intrudes on our reality with malignity. He speaks of the "assaults of chaos" (Lovecraft 1945, p. 15), much like Onishi's initial sense that the insects outside his house are possessed by a "feverish desire to get into [his] house." Onishi reminds himself that this is likely not the case, that neither the insects nor the plants "care about [him] at all" (2020, p. 158), but the weird continues squirming back toward what Miéville (2009) calls the bad-numinous. Although Miéville begins his description of Weird fiction with awe, he quickly shifts toward horror and does not return.

A more wondrously apophatic approach to the weird is not absent in the history of the genre despite a modern preoccupation with horror. Even Lovecraft, most remembered for the eldritch of Cthulhu, offers us the Dreamlands, where "dark symmetries [dazzle] the eye with a beauty more poignant than light," causing the protagonist to "pause in faintness at so much beauty" (2009). Like Lovecraft's Dreamlands, the work of Clark Ashton Smith tends to be passed over in discussions of weird fiction, even though he is generally acknowledged as a key figure in the genre. Although Miéville mentions his name, he makes no further reference to Smith's work in his essay on the genre, a lacuna that is to the detriment of our conception of the eco-weird. Joachim Zemmour offers a compelling analysis of Smith's ability to "conjure up new (ecopoetic) realities" without recourse to the tropes of horror (2021, p. 1). Although many of Smith's stories have strong elements of horror, in stories like "The City of Singing Flame" (2009) he develops a non-horrific weird that borders on religious ecstasy almost directly opposed to Lovecraftian repulsion.

Algernon Blackwood's "The Willows," cited by Onishi (2020) and Germaine (2021) as exemplary of the eco-weird, describes not only the distressing experience "of our utter insignificance before [the] unrestrained power of the elements" but also how this "unknown and immense kind of fear" awakens equally immense experiences of wonder: "the whole magical beauty of it all was so overpowering that a sort of wild yearning woke in me and almost brought a cry up into the throat" (2019, pp. 21, 30). This apophatic tendency goes even further in "The Man Whom the Trees Loved." Here, the cataphatic theological certainty of Mrs. Bittacy recoils from the weirdly wondrous love connecting her husband and the Forest (Blackwood 2019, p. 161). Following her husband into the woods, she experiences what could be described as a dark night of the soul: "Beyond this stealthy silence, just within the edge of it, the things of another world were passing. But she could not know them .... In this time of self-forgetfulness, when she realized that the battle was hopeless, thinking she had lost even her God, she found Him again quite close beside her" (Blackwood 2017, p. 182). Even though she does not understand, from that point forward Mrs. Bittacy stands aside and allows her husband to be taken up into the Entity of the Forest. When we stay with the weird in Weird fiction, it is difficult to distinguish it from the apophaticism of someone like Thomas Merton: "When we seem to be useless and worthy of all contempt, when we seem to have failed … it is in this darkness that we find liberty" (1949, p. 172).

Rebalancing our understanding of the weird back toward apophatic awe is essential if we take seriously, as I do, Onishi's claim that Weird fiction can help us develop our capacity for wonder in ways essential for imagining and realizing ethical futures full of thriving multi-species fields. A conception of the eco-weird based solely on horror can help increase eco-anxiety for the bored, but many of us suffer from an excess rather than a deficiency of eco-anxiety, and the horror dial only goes one way. What is needed, then, is an approach to the weird that offers both the unsayable and the unspeakable. We need both the call to stand before the Weird darkness with fear and trembling and those angels enjoining us not to be afraid. A more apophatically wondrous eco-weird can help us decrease our anxiety enough that we are able to act, but it is often difficult to identify the connection between action in our world and our experiences of Weird wonder in the other worlds of fiction. In the next section, I consider weird TRPGs as a cosmographic technology linking those worlds in a way that allows for wondrous ecological praxis.

8.3 A Cosmographic Praxis of Weird Wonder

Analysis of the eco-weird began by considering film and literature, media that can draw us into experiences of weirdly wondrous worlds but do not offer us much latitude for interactively exploring those worlds outside the vision provided to us by the author or director. True, fans imagine worlds adjacent to yet distinct from the singular world of any given movie, short story, or novel, but this stands in sharp contrast with the swarming proliferation of worlds summoned by games. Alenda Chang (2019) touches on the importance of worlds in her description of ecological games as mesocosms straddling the real and imagined worlds, but the liminal hybridity of these mesocosms is still described in singular terms. Notably, Chang's work focuses on video games and precedes the emergence of the eco-weird from the broader field of critical ecology. Chloe Germaine's ecocritical reading of horror role-playing games (2021) brings Chang's work into dialogue with the eco-weird and, equally importantly, shifts the focus from digital to analog. The result is a vision of tabletop role-playing games as a "generative mode for collaborative worldbuilding" that emphasizes the agency of both players and "the more-than-human beings and agents that emerge from the interworlds generated through play" (Germaine 2021, pp. 129–130).

Germaine offers a useful framework for locating the eco-weird in games, guiding us to consider games in three modalities: as texts that produce representations, as experiences that bridge worlds, and as technologies with generative emergent properties (2021, p. 130). I understand the textual modality as the description of the contents of a given world accessed through the game, the "as-if" statements about the beings, places, and events of the imagined world. In describing games as experiences that "combine aspects of the game world and the world[s] beyond," I understand Germaine to mean that we must pay careful attention to the way of being in the world that we experience as we play. This encompasses our experience of the imagined world, our experience of prime reality, and our experience of their superposition through play. The third modality, considering games as technologies, is especially evocative given Germaine's focus on analog, rather than digital games. My argument that TRPGs are effective vehicles for apophatic Weird wonder builds on Germaine's analysis, but especially on this analog technological mode.

In describing games as technologies, I am not referring to the artifactual elements of games such as rule books and dice, but to the interlocking systems of procedures and techniques. This non-material definition of technology follows usage in meditative traditions, such as Antonio de Nicolâas's analysis (1986) of Ignatian spiritual practice as a technological system. De Nicolâas defines a primary technology as "the habituation in the ordering and repetition of certain mental acts and languages to extend the human sensorium as far as the technology is able to reach" (1986, p. xxi). This approach is consonant with Heidegger's definition of technology as a "mode of revealing" and points out the root term *Technē* refers not only to material crafts but to "the arts of the mind" (1977, p. 13). In this section, I expand this concept into a larger model of games as weird technologies that enhance our capacity for imaginative, perpetually open, cosmographic exploration of possible worlds.

8.3.1 *Multiversal Cosmography*

In my weird cosmographic model of the multiverse, drawing on Mark J. P. Wolf's (2012) model of imaginary and secondary worlds, I conceive of possibility space as a kind of atomic model for thinking about multiversal realities (Fig. 8.1). At the center of the model, like the nucleus of an atom, is what we can call prime reality. The electron cloud of all possible worlds extends, potentially infinitely, outwards from prime reality in every direction of possibility. Just adjacent to my world is the world where my coffee cup is red instead of blue. Further afield is a world where my head is an entire dolphin. Like Dr. Yu Tsun in Borges's "The Garden of Forking Paths," we are surrounded by possibilities, "secret, busy and multiform in other dimensions of time" (1964, p. 42). Here, in the quantum mechanics of multiversal wonder, all possible worlds exist in a state of superposition, perpetually collapsing down from possible futures into inescapable actuality.

Between the infinite reaches of that cloud and the (de)finite concreteness of prime reality exists the sphere containing every world that humans have ever imagined. Presumably, this sphere has continuously expanded since hominids first developed the capacity to imagine, and it will continue expanding for as long as we can imagine. From any given historical

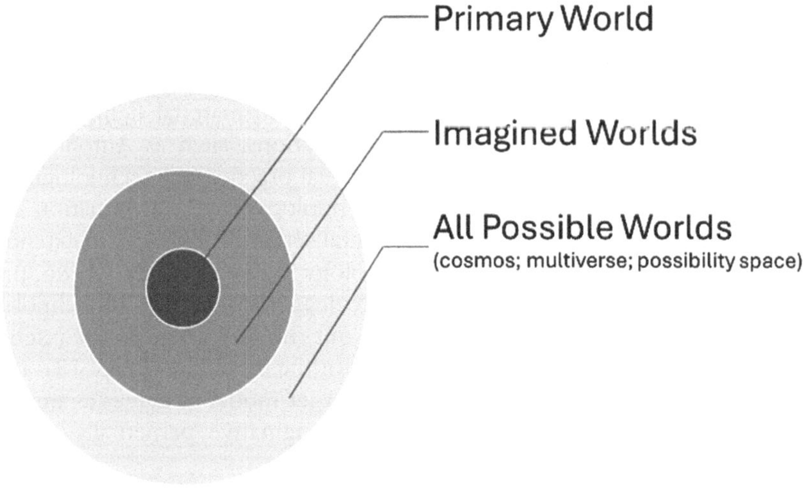

Fig. 8.1 Weird cosmographic model of the multiverse

position, a human can only imagine certain things. Our ability to perceive into the possible has a limit, an imaginative range. It is extremely unlikely that anyone in the Bronze Age imagined a scene like the one in which I find myself writing, on the third floor of a building, causing letters to appear on a flat rectangle in front of me by pressing keys, listening to the sound of music produced years ago by a friend who lives thousands of miles from me. Our imaginative limits, of course, are not only related to material technologies. Any time we invoke the well-worn maxim that it is easier to imagine the end of the world than the end of capitalism, we are also dealing with imaginative range.

Yet our imaginative range is not static. Conceiving of imagination as a mode of perception, specifically that of perceiving some segment of the possible which has not passed into the actual, we can then conceive of technologies of imagination that enhance our perceptive capacity. Just as Marshall McLuhan (2003) argued that the telescope serves as an extension of our eyes and the hammer as an extension of our arms, I argue that we have developed technologies of the imagination that enhance our ability to explore the realms of the possible.

Michael Saler has documented some of this technological development by describing the massive shift in thinking about imagination and other worlds that occurred in the late nineteenth and early twentieth centuries (2012). The development of what Saler calls "the ironic imagination," characterized by playfully treating imagined worlds as if they are real, allowed people to "inhabit the imaginary worlds of fantastic fiction for extended periods of time without losing sight of the real world" (2012, p. 14). I characterize this as imaginative technology, and when combined with the speculative implementation of "realist" technologies of representation like maps and timelines, it coincided with the emergence of a new kind of speculative fiction not previously seen in human history, including the Weird fiction of Lovecraft, Smith, Howard, and the rest. In other words, the technology of ironic imagination allowed both readers and authors of speculative fiction to perceive further afield into the possible.

8.3.2 TRPGs as Cosmographic Technology

The ironic imagination is not the only technology of imagination important in this history. Even more significantly, imaginative technology increasingly integrated elements of play. Lovecraft, Howard, and Smith could be described as playing a kind of game in which they treated each other's worlds as real enough to serve as a foundation for shared imagination, connecting them into a shared mythic universe (Saler 2012). Fritz Leiber's world of Lankhmar emerged as a literal game played with friend Harry Otto Fischer (Leiber 1977). Perhaps the most influential shift in imaginative technology, however, came out of the intersection of Weird fiction with the American Midwest hobby wargaming culture of the late sixties and early seventies, leading to the invention of TRPGs. A brief examination of the core imaginative technologies of TRPGs will help set the stage for my argument that they are especially powerful vehicles for developing eco-weird wonder. Indeed, if we are going to conceive of TRPGs as a kind of technology, then "vehicle" seems to me the most apt metaphor, in that they offer a means of exploring multiversal cosmographic space.

First, TPRGs systematize the collaborative imagination by providing procedures for multiple "cosmographers" to operate the imaginative technology simultaneously, spurring them further into the possible than any of them could venture on their own. This collaborative imagination is not unique to TRPGs. In some ways, it exists any time humans encounter each other's imagination, and improvisational performance of theater and music utilize this technology of simultaneous collaborative imagination in particularly similar ways. Nevertheless, TRPGs advance technologies for collaboratively exploring imagined worlds through systematizing the structural relationships between the imagined world, the narrative, the ludic procedures (i.e., rules), and specialized player roles. Over the past 50 years, games have developed different sets of relationships between the various human and non-human elements of the system, and we could consider each different structural arrangement a basic vehicular model with unique affordances for particular kinds of cosmographic exploration.

The second core imaginative technology of TRPGs is the integration of probability into cosmographic exploration. Random, or at least complexly patterned, phenomena serve as the basis for most kinds of divination that humans have developed, from the *I Ching* to the Tarot. In the nineteenth century, Georg von Reiswitz's development of wargaming set the stage for TRPGs, deploying probability and chance as a means for realistic simulation of war. Significantly, a major goal of *Kriegspiel* was to propel the general-in-training into the uncertainty of the fog of war, which has a sort of diabolic resonance with the cloud of unknowing (Peterson 2012). Reiswitz used dice as a technology for looking into unknown outcomes of actions in an imagined world. TRPGs adopted this usage but also began using chance to "generate" aspects of those imagined worlds, including environments at all scales. If the structural arrangements of human and non-human elements form the architecture of TRPGs' imaginative vehicles, the integration of chance forms the engine, generating powerful non-human constraints that propel the imagination further with less effort.

There is a lot of imagining to be done, and games serve as an assistive technology, offering players both greater endurance and increased cosmographic range. Somewhere in the sphere of possible worlds exist a number of worlds in which humans embrace eco-weird wonder and use it as a guiding virtue to help shape a world of thriving multi-species fields, and the imaginative technology of games can help us discover those worlds. However, as soon as we realize that games give us access to an eco-weird utopia, we also realize, like the residents of Borges' "Library of Babel" (1964), that we have access to an infinite number of futures that appear to be utopia but are in fact diabolical counterfeits.

8.3.3 *Weirdly Wondrous Utopias*

It may seem both naive and dangerous to discuss utopia here. Certainly, the eco-weird was not developed as a utopian concept, and Onishi closes his initial formulation of it by specifically disclaiming the idea that "the eco-weird is a saving mechanism, a panacea for all our environmental problems" (2020, p. 172). This matches a right-minded skepticism toward utopia elsewhere in the eco-critical literature, perhaps most strongly expressed in Haraway's call to turn from "awful or edenic pasts and apocalyptic or salvific futures" (2016, p. 1). And yet it remains an open question whether we *can* fully turn our backs on utopia. When considering eco-weird wonder, Onishi postulates that "when we think about a future world that includes climate disasters, we *must* imagine those futures with flourishing human and nonhuman communities" (2022, p. 35, emphasis mine). Although he is primarily expressing an ethical imperative, I suggest that he also describes a human proclivity toward imagining utopia. Despite Haraway's warning against both despair and hope, it is hard to read the final sentence of *Staying with the Trouble* as anything other than hopeful utopian prophecy: "The Children of Compost would not cease the layered, curious practice of becoming-with others for a habitable, flourishing world" (2016, p. 168). Although superficially this may seem contradictory, I believe it comes into focus if we transpose Haraway's despair and hope onto Onishi's "excess of anxiety" and "deficiency of boredom"

(2022, p. 1). Perhaps the problem with hope is not that it suggests that a better world is possible, but that the closed and overly determined "hope" implied by cataphatic utopian thinking tends toward bored paralysis if not infused with apophatic wonder.

China Miéville argues along these lines, pointing backward to "lifetimes-worth of pre-dreams of New Edens" as suggestive evidence that utopias are necessary, if insufficient and susceptible to totalizing manipulation (2018). The solution, he says, is not to dismiss utopia, but to take its unsayability seriously: "if we take utopia seriously, as a total reshaping, its scale means we can't think it from this side. It's the process of making it that will allow us to do so. It is utopian fidelity that might underpin our refusal to expound it, or any roadmap" (2018). The importance of making here suggests that we have been using "worldbuilding" for the wrong part of the process. In the cosmographic model I've outlined above, we do not make other worlds, we discover them. The building that we do occurs in this world, as our praxis shapes reality. Onishi, seeking an ethical prescription to guide our imagined worlds, one needed because "our imagined futures partially prescribe our actual futures," argues that they should be suffused with eco-weird wonder (2022, p. 34). This is true, but our model of the relationship between imagined and actual worlds must go further if we are to keep our utopias from diffusing into the "faith without works" of imagined worlds as pressure release valves that manage our overwhelming desire to act or condensing into the overly determined cataphatic singularity of so-called utopian movements like eco-fascism (Miéville 2018; Genovese 2023). To avoid these twin errors, we need to endure in a perpetual making, re-making, imagining, and re-imagining. Our entire process of imagining other worlds must be conducted in the perpetually open mode of apophatic wonder. As soon as we imagine a utopia, our process must remind us that we have not described the indescribable utopia, but only our imaginative horizon; from there we must go on imagining. The problem is that we cannot imagine our way out of our imaginative horizon. Only action in the wondrous world, a world that perpetually breaks our categorization of it, can affect this unveiling. Therefore, our model of effective worldbuilding must include a cycling between reflection and action, a process I call the Cosmographer's Circle (Fig. 8.2).

The initial stage of the circle involves reflective observation of prime reality. In observing reality, we identify features of it that seem suboptimal at best and unspeakably horrific at worst. To take a mundane example, as I write, I pause to reflect and realize that I am thirsty and that it is hot in

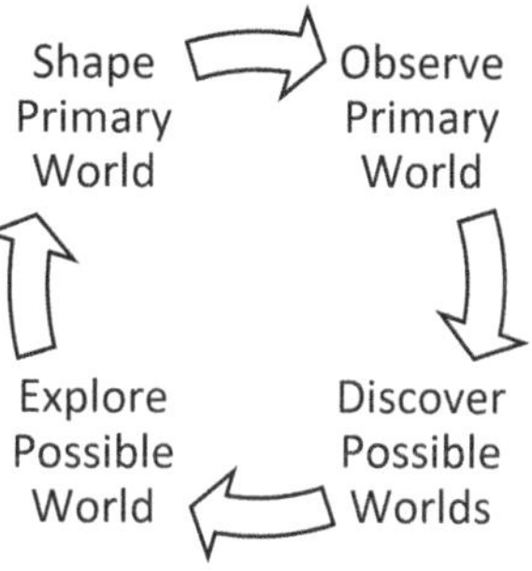

Fig. 8.2 The cosmographer's circle

my office. From here, in the second stage, we gaze out into the possible to identify worlds we think we might like better than this one. I perceive a world where I step away from my office and go buy a cup of coffee. In another, I keep writing and only pause momentarily to take a sip from my water bottle. Moving beyond perceiving an imagined world, in the third stage we enter into it, performing what Edward Casey (1976) calls "imagining that." Imagining the coffee shop world, I find that in this world I enjoy the cool spring air but have less money, take longer to write my chapter, and end my day rather dehydrated. The water bottle world, however, seems quite nice, if a bit hot, so I modify it slightly to include opening the window. In this stage of imagining-that, I alter my consciousness, if only slightly in this instance. I have remembered that the window opens, that the weather is pleasant, and that I've probably already had enough coffee today. At this point in the Cosmographer's Circle, I am like Campbell's hero, having discovered the boon of water and fresh air in the underworld of my imagination, but like that oft-maligned Eurocentric phantom, I too must decide to return to prime reality with the boon of my new perception. A better world is possible, and I transmute the reflection into action as I pause mid-sentence to drink and open. Having done that, I begin the cosmographic cycle anew, finding that my previous actions failed to bring about a perfect utopia. I realize something I hadn't before: my back is sore, and my office is a mess. Following the cycle, I stretch, throw some things away, and sit back down to write.

While I have used a small example here for illustrative purposes, we also use this basic cycle of reflection and action in the world at larger scales, all the way up to the scale of our reaction to ecological crisis. Games can assist us at each stage of this process. Although TRPGs have not historically focused on observation and praxis, Augusto Boal's argument that improvisational theater games can serve as rehearsal for revolution (2019) could serve as a useful starting point for future consideration of how TRPGs can similarly help players continue moving around the Cosmographer's Circle. Here and in the case study that follows, however, I have limited myself to the more developed ways that TRPGs assist us in discovering and exploring weirdly wondrous worlds.

8.4 Weirdhope

The weird has been a key piece of the TRPG landscape from the beginning. Lovecraft, Dunsany, and Howard all feature in Gary Gygax's list of reading that inspired *Dungeons & Dragons* (1979), and the Lovecraftian horror RPG *Call of Cthulhu* regularly holds a solid second place in ranking of popular role-playing games like Roll20.net's *Orr Report* (Roll20.net 2021). Beyond this general weird influence in TRPGs, and even beyond Germaine's (2021) list of ecohorror games, recent years have brought a growing number of games that could be described as eco-weird. *Dream Askew* (Alder and Rosenbaum 2019), although not intensely ecological, "explores the story of a queer enclave amid the collapse of civilization" in a setting that includes a psychic maelstrom, mutant plants, and post-apocalyptic scarcity; *Spirit & Tech* (MacAdams 2021) sends characters into "nightmarish spaces, where all of the old word's wrongs have become awful, twisted physical manifestations"; *Moon Elves* (Maik 2021) imagines a lunar future of "mutualistic-symbiotic integration" between post-humans, telepathic trees, mycorrhiza, and microscopic life; and *I am The Forest* (Faerber 2022) is a solo RPG that could be described as playing "The Man Whom the Trees Loved" from the perspective of the trees. The work of artist and game designer David Blandy stands out in this growing field of eco-weird games, and in this final section, I will consider three of Blandy's games, *The World After* (Blandy and Goulson 2019), *Lost Eons* (Prentice et al. 2021), and *Eco-Mofos* (Blandy 2024), as examples of games that not only represent eco-weird worlds but effectively deploy cosmographic technology to encourage experiences of those worlds suffused with apophatically tinged wonder.

Eco-Mofos, as described on its Kickstarter page, is "a mid-future eco-punk ruin-delving survival game, as player characters (PCs or Punks) seek a safe homestead to start a new community, Weirdhope not Grimdark" (Locke and Blandy n.d.). Set 200 years after "everything went to frak," in ecological catastrophe, the game's cosmography shares many features with two of Blandy's previous games, *Lost Eons* and *The World After*. All three guide players to worlds in which human, non-human, and post-human life find ways to thrive in the time after climate cataclysm. Although not connected in a strictly canonical single timeline, they share enough features to be considered a kind of trilogy describing adjacent regions of the multiverse. For the sake of having a name for the entire cosmographic territory, I will refer to this "Blandyverse" as Fain, drawing on the title for the world used in *The World After*.

Eco-Mofos lays out the basic form of Fain most clearly, explaining that "the wealthy, who created this mess, fraked off to Mars while the lucky ones hid in vast underground bunkers. We were left to fend for ourselves on the surface, eking out an existence in the broken ruins they left behind, searching for remnants of a past world to help us survive, searching for a place to call home" (Blandy 2024, p. 8). Although the wealthiest leave Earth for Mars, they return to Fain periodically as the Cryorg, always looking for an opportunity to re-establish Capitalist control of the world. The cataclysm also sets off a re-enchantment of the world by means of an Essence, which *The World After* describes as "substance that is secreted from the Earth as a multibody when it experiences catastrophic harm" (Blandy and Goulson 2019, p. 8). Essence, "flows across the surface of the World, linking all living things," stimulates rapid evolutionary adaptations and "unlock[s] the sentience of many elemental forces" (Blandy and Goulson 2019, p. 8). The representational text of these worlds is eco-weird through and through, from "sentient reed bed[s] that you can hear whispering on the wind" to the post-human Uroch, humanoid cave-dwellers interpenetrated with symbiotic fungus (Prentice et al. 2021, p. 7).

Rather than focusing on the eco-weirdness of Blandy's games as representational texts, however, I would like to consider their cosmographic technologies. Part of what makes them such interesting case studies is that the similarities between the games' worlds help bring the differences in cosmographic technology into focus, allowing us to see some technologies as weirder than others. The cosmographic technologies of the latter two games, *Lost Eons* and *Eco-Mofos*, are particularly supportive of apophatic eco-weird wonder when compared with *The World After*. In order not to give the impression that Blandy innovated these designs whole cloth it should be noted that both games cite a number of design influences, most notably Chris McDowall's *Into the Odd* (2022). Although *Lost Eons* and *Eco-Mofos* have much in their designs worth considering, it is their apophatic anti-canonicity and oracular procedurality that contribute most powerfully to their weirdly wondrous cosmography.

The World After follows a traditional "worldbuilding" approach to Fain, offering a detailed timeline and map of the world along with pages of ethnographic information about various societies. While the resulting text richly represents a world dense with eco-weird wonder, as a piece of cosmographic technology, it falls short in one essential way. Whereas eco-weird wonder remains perpetually open to possibility, *The World After*'s cataphatic, canonical technology condenses the quantum cosmographic field into a single world, presented as eminently knowable even without direct experience. Admittedly, if we compare this to a non-ludic cosmographic artifact like "The Willows," the technology of the game offers an almost infinite increase in the wondrous superposition of possible worlds. While various readers might imagine physical specificities of a character or scene differently and thus envision slightly different worlds, every group of players visiting canonical Fain will visit a slightly different version of the world, either because a group might modify or ignore some pieces of the canonical lore or simply because each group will make different decisions in the world once they begin playing. Nevertheless, the establishment of a canonical Fain definitionally presents even the weird world of Fain as knowable. This can be seen more clearly by way of contrast with the anti-canonical approach of *Lost Eons* and *Eco-Mofos*.

"LORE IS NOT CANON," asserts *Lost Eons* in large text. "Lore is just a starting point …. Imagine a world after capitalism" (Blandy and O'Connel 2021, p. 3). *Eco-Mofos* takes this decentering of cataphatic canon even further, on a page headed "WHAT WE KNOW," explaining, "The world is created at your table, and that's the only reality that matters" (Blandy 2024, p. 12). Similarly, the latter game simplifies *TWA*'s 20+ pages of detailed dogma into 21 enigmatic statements that suggest a cosmographic region rather than define a particular world. Statements like "Things that were have grown and twisted, and new horrors have flourished" raise more questions than they answer. With scalpel-like precision, they incise the borders of a cosmographic region, but their apophatic silence holds the adjacent worlds of that region open in weirdly wondrous superposition. While a game like TWA might cataphatically describe a weirdly wondrous world, the apophatic technology of anti-canonical games allows them to maintain the Weird wonder of the multiverse, "the open and generative possibility that stands at the superposition of relation before collapse" (Onishi 2023, p. 121).

Although notable, Blandy's anti-canonical approach is not unique, even among eco-weird games. Germaine (2023) has noted that in *Hollows of Desolation* (Murphy 2021) the world is almost entirely generated through play, and similar patterns exist in *Dream Askew*. I argue that this commonality exists, in part, because as a cosmographic technology, anti-canon is itself inherently more open to Weird wonder than cataphatic canon. Although aspects of anti-canonicity can be seen throughout the last half-century of TRPG design, the term "anti-canon" seems to have been coined by Luka Rejec, whose dying earth *Ultraviolet Grasslands* RPG borders on the eco-weird. Rejec explains his anti-canonical practice in terms that resonate with both the cosmographer's circle and Weird wonder, rejecting "a single *right* way of being" and trying to ensure that the world generated through play "is a unique and ephemeral creation that appears for a moment in the collective imagination of the group, before dissipating again, falling to the subconscious to fertilize new ideas, new worlds, new stories in some future time" (2019).

Oracular procedurality, although not necessary for anti-canon, complements it nicely as a particularly powerful form of emergent narrative. Emergent narrative design does not require the gamemaster to prepare a scenario or narrative in advance. Instead, the narrative emerges as players "play to find out what happens" (Baker and Baker 2016). In many cases, this emergence depends primarily on the improvisational skills of the gamemaster to fill in details of a general structure. *They are Hollows of Desolation*, for example, tells the gamemaster to identify a series of sets, or narrative units, based on the collaboratively built environment. Sets are connected with particular locations, and each set has a set goal, "what the players are currently trying to accomplish" (Murphy 2021, p. 11). In other games, this emergence depends primarily on the improvisational conversations between players. *Dream Askew*, for example, follows a "no dice, no masters" philosophy and generates scenarios without a gamemaster through a technology called "Idle Dreaming," which encourages players to muse together about "whatever is interesting, or unknown, or scary, or beautiful about this place that you're building together" (Alder and Rosenbaum 2019, p. 24). This discussion leads players to an idea about a scene they would like to play through. If a scene's ending leaves the group in doubt about the next scene, they return to Idle Dreaming. Whether using a gamemaster or not, these structure-and-improvisation technologies promote wondrously emergent worlds, and my personal experience has found that what emerges is often weirder than what any individual is likely to have developed in advance. Further, because the players jointly experience the contingent openness of the events, the result is more likely to lead to wondrous awe when everything comes together in a sublime moment.

Whereas *Dream Askew* eschews dice altogether, *Eco-Mofos* and *Lost Eons* lean heavily on their generative potential. The technology in both games is similar but finds a more elegant expression in *Eco-Mofos*, which describes it as "procedural adventuring" (Blandy 2024, p. 92). The basic procedure has characters explore a series of pre-keyed "point crawl" maps that abstract their exploration at three scales named after the rough amount of time it would take to cross from one side of the map to the other. At each point on the map, the GM uses dice and tables to generate the general environment, a unique feature, and possible encounters. A section of generated wilderness, for example, might present "a dark forest, vines reaching and recoiling" in which "the floor is a carpet of moving centipedes" and a "fungal troll," "dormant, regenerating" (Blandy 2024, pp. 108, 113). The solo rules of the game take procedural adventuring a step further with the O.R.A.C.L.E. system, expanding a now common game design term for random generation systems into an acronym for a 5-stage process: "Observe your environment; Resolve goals and obstacles; Act taking consequences or rolling … Conclude what the impact is; Leave evidence … writing in your journal" (Blandy 2024, p. 160). In the Observe stage, a player curious about the centipedes might roll on the Portent table to generate a Delphic utterance like "foolish/child/victim; pain/wound/infection." Does this mean that the troll has fallen victim to the centipedes and is attempting to regenerate even as they perpetually wound him? However the player interprets the results, Blandy describes the philosophy behind them in his advice to the players: "The game world is organic, malleable and random. Use random tables and generators to develop situations, not stories or plots …. From situations … stories of wonder and resistance will emerge" (2024, pp. 14–15). Through this cosmographic technology, *Eco-Mofos* encourages a wondrous orientation between the player, game artifacts like dice and tables, and what Germaine calls "the more-than-human beings and agents that emerge" through their weird loopings.

Having considered Blandy's games as both texts and technologies, I close by considering their experiential aspects, the mode of dwelling in and between worlds they encourage. Blandy describes the experiential goal of *Eco-Mofos* as "weirdhope" which is remarkably similar to Onishi's "eco-weird wonder." Like Weird wonder, Blandy seeks a middle ground between anxiety and boredom, urging the player to "lean into the Weirdhope. Things can be scary and odd, and damn, bad things happen, but things are getting better" (2024, p. 8). Like Weird wonder, weird-hope is a virtuous capacity connected with ethical praxis in the face of unspeakable horrors. The safety section warns players that "This game contains a lot of stuff that could be difficult to deal with including addiction, gore, body-horror, cybernetics, environmental breakdown (and recovery), non-human consciousness, cannibalism, dead bodies (including children), corporate greed, heinous billionaires and selfish small-mindedness," but ends with an admonition: "It's a game about weird hope, so let's frak stuff up and make things better" (2024, p. 17). Similarly, *Lone Eons* opens with a synopsis suffused with apophatic Weird wonder: "You will emerge from the darkness into a new light, one dominated by inscrutable and terrible forces you must seek to understand. Using these pages create a post-human character, ready to face the mysteries and dangers of the far future" (2021, p. 3). There is a beautiful ambiguity to the grammar here. The immediate reading is that the player is admonished to create a fictional character for the game's imagined future. Read esoterically, the text instructs players to use the game as a tool for shaping their own character, developing virtues that leave them better equipped to face the mysteries and dangers of the very real future.

8.5 Conclusion

The world is ending, and that is weird. The world has always been ending, which is weirder. Weirdest of all, the world is now ending in a way that it has never been ending before, and it will continue to go on ending in ever-weirder ways that unfold as far into the future as humanity can imagine. In this ever-ending, never-ending weirdness, weird seeds of new life perpetually emerge from the decayed remnants of unspeakably ancient beings, human and otherwise.

Games can help us stay with the weird, to embrace the discomfort of it rather than domesticating, fearing, or ignoring it. As emergent representational texts, games help us stay with weird wonder in the world, following with wondrous eyes each twist and turn of its unfolding, even when those twists and turns present unspeakable horrors. As assistive cosmographic technologies, games lend us endurance in staying with the weird in our imaginations, holding to an open superposition of futures both possible and impossible and rejecting the boredom of false hope that believes it can reduce the unsayable to a singular utopia. Most importantly, as experiential practices that connect and shape worlds, we can stay with weird wonder as we act in our own world. We must, as Onishi says, "fumble our way in the dark world of material mystery" (2023, p. vii). Weird wonder suggests that this may not be the darkness of hell but "the truly mysterious darkness of unknowing" (Sells 1994, p. 35). In Genovese's poetic words, "may we head toward that faint gleam in our misty future—that subterranean light buried beneath the horizon which promises nothing to us except the possibility of something radically different and darkly beautiful, if only we might wield our eternal imaginations to materially enact its spectral desires" (2023, p. 37).

References

Alder, Avery, and Benjamin Rosenbaum. 2019. *Dream Askew & Dream Apart*. United States: Buried Without Ceremony.

Baker, D. Vincent, and Maguey Baker. 2016. *Apocalypse World*. 2nd ed. Gamefield, MA: Lumpley Games.

Blackwood, Algernon. 2017. *Ancient Sorceries and Other Weird Stories*. CreateSpace Independent Publishing Platform.

Blandy, David. 2024. *Eco mofos*. Playtest Booklet 1.0. https://davidblandy.itch.io/eco-mofos. Accessed 16 May 2024.

Blandy, David, and Laurie O'Connel. 2021. *Lost Eons*. Itch.io. https://davidblandy.itch.io/lost-eons.

Blandy, David, and Matt Goulson. 2019. *The World After: Roleplaying Post-Humanity as They Forge a New World after the Climate Cataclysm*. Southend on-Sea, [England]: Focal Point Gallery.

Boal, Augusto. 2019. *Theatre of the Oppressed*. Translated by A. Charles, Maria-Odilia Leal McBride, and Emily Fryer. London: Pluto Press.

Borges, Jorge Luis. 1964. In *Labyrinths: Selected Stories & Other Writings*, ed. Donald A. Yates and James E. Irby. New York: New Directions.

Casey, Edward S. 1976. *Imagining: A Phenomenological Study*. Studies in Phenomenology and Existential Philosophy. Bloomington: Indiana University Press.

Chang, Alenda Y. 2019. *Playing Nature: Ecology in Video Games*. Electronic Mediations 58. Minneapolis: University of Minnesota Press.

De Nicolâas, Antonio T. 1986. *Ignatius De Loyola, Powers of Imagining: A Philosophical Hermeneutic of Imagining Through the Collected Works of Ignatius De Loyola, With a Translation of These Works*. Albany: State University of New York Press.

Faerber, Seth A. 2022. *I Am the Forest*. Itch.io. https://grismund.itch.io/iamthe-forest. Accessed 16 May 2024.

Genovese, Taylor R. 2023. *Dark Cosmism: Or, the Apophatic Specter of Russo-Soviet Techno-Utopianism*. Arizona State University.

Germaine, Chloe. 2021. Dark Forests and Doomed Adventurers: An Ecocritical Reading of Horror Roleplaying Games. In *Generation Analog: The Tabletop Games and Education Conference*. Online: ETC Press.

Gygax, Gary. 1979. *Dungeon Master's Guide*. Lake Geneva, WI: TSR Games.

Haraway, Donna Jeanne. 2016. *Staying with the Trouble: Making Kin in the Chthulucene*. In *Experimental Futures: Technological Lives, Scientific Arts, Anthropological Voices*. Durham: Duke University Press.

Heidegger, Martin. 1977. The Question Concerning Technology. In *The Question Concerning Technology and Other Essays*, trans. William Lovitt. New York: Garland Publishing.

Kant, Immanuel. 1991. *Observations on the Feeling of the Beautiful and the Sublime*. Translated by J.T. Goldthwait. Berkeley: University of California Press.

Leiber, Fritz. 1977. *Swords and Deviltry. The Fafhrd and the Gray Mouser Saga of Fritz Leiber*. Boston: Gregg Press.

Locke, Daniel, and David Blandy. n.d. eco mofos!! Kickstarter page. https://www.kickstarter.com/projects/daniellocke/eco-mofos. Accessed 16 May 2024.

Lovecraft, H. P. 1945. *Supernatural Horror in Literature*. New York: B. Abramson.

Lovecraft, Howard Philips. 2019. The Dream-Quest of Unknown Kadath. https://www.hplovecraft.com/writings/texts/fiction/dq.aspx. Accessed 16 May 2024.

MacAdam, Marren J.T. 2021. *Spirit: Tech: The Scars of Our Rule*. Itch.io. https://marrensmusings.itch.io/spirit-tech. Accessed 16 May 2024.

Maik. 2021. *Moon elves*. Itch.io. https://maik-malaik.itch.io/moon-elves. Accessed 16 May 2024.

McDowall, Chris. 2022. *Into the Odd. Remastered*. Manchester, England: Bastionland Press.

McKenna, Terence. 1998. Terence McKenna's Final Earthbound Interview. By John Hazard. *Reality Sandwich*. https://www.youtube.com/watch?v=GdEKhIk-8Gg. Accessed 3 June 2024.

McLuhan, Marshall. 2003. *Understanding Media: The Extensions of Man*. Edited by W. Terrence Gordon. Critical ed. Corte Madera, CA: Gingko Press.

Merton, Thomas. 1949. *Seeds of Contemplation*. New York: New Directions Publishing.

Míeville, China. 2009. Weird Fiction. In *The Routledge Companion to Science Fiction*, eds. Mark Bould, Andrew M. Butler, Adam Roberts, and Sherryl Vint, 510–515. Routledge Companions. London: Routledge.

———. 2018. The Limits of Utopia. Climate & Capitalism. March 2. https://climateandcapitalism.com/2018/03/02/china-mieville-the-limits-of-utopia/. Accessed 16 May 2024.

———. 2019. Silence In Debris: Towards an Apophatic Marxism—China Miéville. *Salvage* (blog). April 2. https://salvage.zone/silence-in-debris-towards-an-apophatic-marxism/. Accessed 16 May 2024.

Murphy, Gordie. 2021. *They Are Hollows of Desolation*. Itch.io. https://gmurphy.itch.io/. Accessed 16 May 2024.

Onishi, Brian Hisao. 2020. Weird Ecologies and the Uncanny in The Happening. In *Philosophy, Film, and the Dark Side of Interdependence*, ed. Jonathan Beever, 157–174. Lanham: Lexington Books.

———. 2022. Weird Environmental Ethics: The Virtue of Wonder and the Rise of Eco-Anxiety. *SATS* 23 (1): 33–53. https://doi.org/10.1515/sats-2021-0016.

———. 2023. *Weird Wonder in Merleau-Ponty, Object-Oriented Ontology, and New Materialism*. Cham: Springer Nature Switzerland. https://doi.org/10.1007/978-3-031-48027-0.

Otto, Rudolf. 1923. *The Idea of the Holy*. http://archive.org/details/in.ernet.dli.2015.262513.

Peterson, Jon. 2012. *Playing at the World: A History of Simulating Wars, People and Fantastic Adventure, from Chess to Role-Playing Games*. San Diego, CA: Unreason Press.

Prentice, Andy, David Blandy, and Laurie O'Connel. 2021. *Lost eons*. Itch.io. https://davidblandy.itch.io/lost-eons.

Rejec, Luka. 2019. Anti Canon Worlds and the UVG. *Wizard Thief Fighter* (blog). March 5, 2019. https://www.wizardthieffighter.com/2019/anti-canon-worlds-and-the-uvg/.

Roll20.net. 2021. The Orr Report Q3 2021. *Roll20 Blog*. November 11. https://blog.roll20.net/posts/the-orr-report-q3-2021/.

Saler, Michael T. 2012. *As If: Modern Enchantment and the Literary Pre-History of Virtual Reality*. New York: Oxford University Press.

Sells, Michael A. 1994. *Mystical Languages of Unsaying*. Chicago: University of Chicago Press.

Smith, Andrew, and William Hughes, eds. 2013. *EcoGothic*. International Gothic. Manchester: Manchester University Press.

Smith, Clark Ashton. 2009. The City of Singing Flame. In *The Return of the Sorcerer: [The Best of Clark Ashton Smith]*, ed. Clark Ashton Smith and Robert E. Weinberg. Rockville, MD: Prime Books.

Wolf, Mark J.P. 2012. *Building Imaginary Worlds: The Theory and History of Subcreation*. New York: Routledge.

Zemmour, Joachim. 2021. (Wo)Men-Flowers and Urban Gardens in the Weird Fiction of Clark Ashton Smith: Vegetal Hypallage as an Ecopoetic Device. *Textes et Contextes*, no. 16–1 (July).

CHAPTER 9

Tabletop Eco-Weird: Gameplay Experience and Ecological Ethics

Chloé Germaine

9.1 Introduction: Eco-Weird as Experience

Tabletop games are eco-weird experiences. Moreover, games demonstrate that the eco-weird is best understood as an aesthetic, a phenomenological modality comprising sensory and affective experiences. At the heart of the eco-weird experience is a 'wound' that some ecological philosophers have called 'nature'. For Timothy Morton (2017), 'nature' names the ragged site of coexistence marred by an attempted 'severing' of human being from ecological being. For Toadvine (2012), nature is the radical alterity from which human experience originates and which disrupts human 'enjoyment' of the world from within that experience. The word 'wound' comes into play here because Toadvine explores the ambiguity of the human's paradoxical separation from and embeddedness within nature as a kind of 'rupture' that is the distance between human beings and the

C. Germaine (✉)
The Manchester Game Centre, Manchester Metropolitan University, Manchester, UK
e-mail: C.Germaine@mmu.ac.uk

B. H. Onishi, N. M. Bell (eds.), *The Call of the Eco-Weird in Fiction, Films, and Games*,
https://doi.org/10.1007/978-3-031-77126-2_9

elements on which they depend (2012, p. 173). For Toadvine, the human-nature relation is not a pure separation (a divide or severing) nor should it be understood as a point of fusion, but, rather, through the 'principle of dehiscence' (2012, p. 188). The eco-weird experience is of nature as this dehiscence, and its expression in human experience, a process of splitting and suturing that plays out over and over. This is the 'human-nature' relationship, then; it is neither a dialectic nor a binary, but a mutual becoming that is damaged and damaging. To encounter the eco-weird through games is to experience a paradoxically estranging intimacy with the environment that is one version of ecological belonging on a damaged planet. In this chapter, I will expand on this description of the eco-weird through a discussion of the affordances of games more broadly and through a case study on the specific board game, *Sleeping Gods*, designed by Ryan Laukat (2021). I will show that games are weird, even if they do not seem to be about anything weird, such as tentacle monsters or eldritch horrors. In fact, it is very important not to equate the eco-weird with these generic features. Games are weird not because they are about Cthulhu, but because gameplay is an entanglement of more-than-human subjectivities and relationships, and makes plain the fallacy of human exceptionalism and agency upon which our perceived separation from 'nature' is founded. Tabletop gameplay is especially weird, I will claim, because it occurs across connected but discontinuous sites: the game board, the tabletop, the imagination. Games across the tabletop require a continual shifting of human players' positionality among the more-than-humans with whom they play. Through my case study, I consider tabletop gameplay as a mode of co-worlding that remains provisional, open, and generative in its potential to bring forth as-yet unrealised possibilities from the virtual into the actual.

9.2 Games as Speculative Fabulation

Tabletop games are a mode of speculative fabulation in the sense described by Donna Haraway (2016). For Haraway, 'SF', defined variously as 'science fiction', 'speculative fabulation', or 'string figuring', is 'the patterning of possible worlds and possible times' (2016, p. 31). Games enact patterns that have the potential to disrupt patterns, which is to say that they are a form of SF that allows players to reconfigure the patterns, or structures, of familiar, everyday life. This is because games are co-created performances, or phenomena, that conjure temporary worlds from myriad materials. Through the co-creation of temporary (game)worlds, players propose and experiment with alternative forms of being and relating, and with different ways of *dwelling*, a word I use throughout this chapter to describe 'becoming with' the environment, deliberately echoing its use in philosophy and elsewhere to denote 'staying within', 'housing', and the home—these 'homely' aspects of subjectivity are precisely what the play of board games disrupts, despite its construction in the popular imagination as a cosy, homely activity.

Games do the work of speculative fabulation in varied ways. In this chapter, I am using the term 'tabletop games' to encompass a range of gaming media. These include pen, paper, and dice roleplaying games, or RPGs, (the most famous example of which is the fantasy RPG, *Dungeons & Dragons*), card, and board games (including the hobby games made popular by Games Workshop, which organise play within fantasy worlds represented on the board; war games, which tend to simulate historical combat; and euro-style board games, which avoid direct conflict). There are also hybrid games that employ analogue and digital components, with play taking place through screen media and across the tabletop. These types of games suggest the different experiences gameplay entails: conflict and combat with other (human) players or against the game itself; temporarily inhabiting another world by engaging with its representation on a game board, or through verbal descriptions; touching, holding, and moving playing pieces; rolling dice to decide what happens next or to resolve conflict; writing; drawing; talking and listening. These multi-sensory experiences are facilitated by the participation of more-than-human beings, including wooden pieces, paper, and card components, flat (2D) representations, 3D miniatures representing characters and monsters, as well as the imagined, but nonetheless *agentially* real, characters, worlds, and environments that come into being through play.

Because games require simultaneous cognitive attention and embodied physical play, their more-than-human participants continually manifest in the foreground of experience. This is particularly apparent in tabletop games because they do not use so-called 'immersive' technologies, such as headsets, ergonomic controllers, or HD screens, which are designed to smooth over the mediation from the world of the player to a digital world within a video game. The facilitators of immersion in tabletop games are many, heterogeneous, and discontinuous—lots of bits and bobs. They are not bounded within a digital space: they must be unboxed, punched out, picked up, put down, moved across a board, shuffled, drawn, thrown, and so on. It is not only that board game bits and bobs draw attention to the more-than-human networks that facilitate play, gesturing to processes of extraction, production, and consumption, to more-than-human habitats via the organic and inorganic materials from which they are made, but that these bits and bobs *actually create* immersive game realities despite their heterogeneous nature. In studies of sensory immersion in gaming, immersive experience is judged via the 'high realism' offered by digital game environments, by the 'level of fidelity' technology can provide in mediating players' experiences (Farkas et al. 2020, p. 59). Reality, however, is not smooth, nor is it loyal to human perceptions. Therefore, board games are no less real and no less immersive than video games by virtue of their discontinuous materiality. Indeed, as Paul Wake (2019) argues, the tactile mechanics of board games facilitate immersive gameplay. Through my case study, I expand on this claim by suggesting that the immersion provided by board games through their heterogeneous materials is of a different type to that offered in video games. In part, board game immersion comes from running the game mechanics, which require human players to do the absorbing work of making the game (unlike in video games, where the black box of the code does most of the work). As in 'puzzle-based immersion' (Ermi and Mäyra 2005), players can become absorbed not only in the storyworld but in running the game. Such absorption, in eco-weird gaming experiences, might entail forgetting the self, with the human player becoming one of the more-than-human component parts of the game, an automata, a runner of sub-routines. Certainly, board game play involves a suspension of the self in the service of others (either actual human guests whom one is hosting, or being hosted by, or the more than human components on whose behalf the players act). This mode of game immersion engages players with, rather than seeks to smooth over, the messy business of co-worlding.

All players of the game, from human participants to component parts, are co-creators of its world, co-enactors of its speculative fabulation: in *Forests of Pangaia* (2023), an action-economy board game, for example, wooden tokens pretend to be trees, asking players to move them into position on the board, while human players pretend to be forest spirits, drawing cards and making decisions based on these draws that will affect the patterns made by the trees. The patterns that emerge through the more-than-human performances of tabletop games are only temporary; they are provisional allegiances, re-negotiated and re-assembled during each play session. Forests grow and are felled, replanted anew each time the board is set out and the card deck shuffled. Gaming is thus a site for the 'passing

[of] patterns back and forth, [of] giving and receiving', actions that are integral to speculative fabulation (Haraway 2016, p. 12). The tabletop is a site of intimate exchange, of imagining together in what turn out to be weird ways.

The imaginative component of tabletop gaming, which entails bringing forth worlds, entities, characters, histories, and objects that do not exist in the 'real' world, also makes clear that speculative fabulation requires the virtual as well as the actual. The virtual designates what is not in actuality but what could be. The virtual names the possibilities for being otherwise that inhere in immaterial and material conditions. Although these possibilities are not actualised, they already exist within the bounds of what is real. Therefore, the virtual is not an antonym for the real. Games are a good way of thinking through this proposition: they have already been described as 'half real' by Jesper Juul (2005). Juul argues that the aesthetic experiences of games hinge on the interaction of 'real' rules (the mechanics that structure play) and imagined, fictional worlds (that are created through play). His designation of games as 'half real' is evocative of the conception of the virtual, but thinking of this in weird ways takes us beyond the implications of Juul's formulation. Games do not simply straddle the real and the imagined, paltry half-and-half things; they provide a passage between the virtual and actual. The realm of Ravenloft, which adventurers explore in *Dungeons and Dragons*, for example, is brought *into* reality through drawn maps, plastic miniatures, and shared words across the tabletops. In Jervis Johnson's board game, *Blood Bowl* (1986), to give another example, when an Orc blitzer scores a touchdown, she is rendered vividly real (in the form of a painted metal or plastic figure), animated by the hands of her human 'coach' moving her across the squares of the board, these movements determined by the mathematical calculations in the game's rules and the roll of a dice, and she is given vibrancy and depth by the lore of the Warhammer storyworld in which the game is situated. As these examples show, the virtual dimensions of tabletop gaming work across its imagined and material components. Games provide a good case study of Gilles Deleuze's philosophy of the actual and the virtual, then, which describes the virtual as residing within materiality as forms not (yet) actual but nonetheless real. More than a synonym for the 'possible', the virtual is 'fully real', a realm of reality in which all objects partially exist, and from which actual objects, relations, and situations arise (Deleuze 2014, p. 272). Similarly rejecting a possible/real binary, Maurice Merleau-Ponty suggests the virtual is a variant of the real, a sphere 'vaster than the particular case of this world' (2007, p. 417). I imagine objections being raised to my choice of a football-playing orc as an example of Deleuze and Merleau-Ponty's ideas since she could never *really* be an instance of the real, however I stretched that category. Nonetheless, what I want to demonstrate is a weird truth of gaming: the orc is already real, though her existence does require an expansion of local realist and reductive materialist definitions of reality.

There is an ethical proposition entailed in thinking of games as partaking of the virtual, as providing a passage between virtual and actual. Not only does this move beyond an impoverished idea of game experiences as 'half' real but also allows for thinking through the futural dimension of games. The virtual in games comprises possible forms that do not exist but that could give rise to new, transformed conditions of reality, as well as already existing forms and ideas, the congealing of strange combinations and phenomena that are the province of the eco-weird. Tabletop games facilitate this through their structural openness. Each game performance reassembles the multiple participants involved in play, remaking the world anew each time, giving rise to opportunities for new combinations, new relationships, and new modes of encounter. The rules of the game may remain the same, as do the physical pieces inside the box, but the experience of play is different each time. In part, this is because of the uncertainty built into games, because of their probabilistic rather than deterministic mechanisms. Situations unfold according to the whim of a dice roll or that of a random card draw. Such probabilistic mechanics function as the passage between the virtual and the actual, leading to new combinations, unplanned scenarios, and, often, the need for human players to reorient themselves before the next turn. This is the *extramaterialist* dimension of games and one that connects gameplay to the non-deterministic nature of worlding to the openness of the future to a 'becoming that enables divergence from what exists' (Grosz 2004, p. 184). This is the case even when the materials assembled remain unchanged: the 'movement of virtuality' does not depend on the injection of new materials but, rather, shows that unpredictability is a condition of material existence, that it inheres within the materials already at hand and that the past does not exert deterministic force over the future (Grosz 2004, p. 8). Contrary to a determinist view of games as closed systems, then, games understood as eco-weird experiences unleash what Grosz calls the force of the future and the overwriting or restructuring of present conditions by as-yet unrealised futures. The speculative fabulation of gaming that readily engages players with weird outgrowths from the virtual activates the future dimension of eco-weird ethics, then, the imperative to imagine the world otherwise than it currently is, to bring about situations that are discontinuous with past and present, to find in the vast store of virtual forms surprising, and untimely, combinations that will overcome the 'game over' scenarios that dominate cultural and political climate imaginaries.

9.3 On Why Games Just *Are* Weird

Underpinning my claim that games are eco-weird is a definition of the eco-weird that turns on the contradiction of ecological belonging in a time of climate crisis. I develop this definition in my book on children's literature, where it emerges through dialogue with fiction and with theories of the uncanny, the planetary, and the gothic (Germaine 2023). I identify spaces evoked in fantastic novels, particularly oceans, as ecological sites that function simultaneously as intimately familiar homes and estranging alien realms for their inhabitants, arguing that it is this 'contradiction that encapsulates the ecoweird' (2023, p. 135). I also argue that the eco-weird in literature exposes the precarious ethical relations involved in ecological intimacy, in being at home, or dwelling, in the world, but that these revelations never collapse into ecophobic representations (Germaine 2023, p. 125). In this chapter, I extend my discussion of the eco-weird to games, arguing that gaming, perhaps even more so than literature, can reveal the wounds at the heart of human experiences of the world (Germaine 2023, p. 126).

As the words 'experience' and 'wounds' suggest, the eco-weird is a phenomenological mode to do with human situatedness within, and the emergence of human perception from, a more-than-human world. The eco-weird is an aesthetic that has the following features: (1) ontological defamiliarisation, prompting feelings we might identify as uncanny; (2) a paradoxical intimacy and estrangement, of being at home in, yet alienated from, a more-than-human world; (3) a disclosure of the interdependencies with, and obligations to, other critters than ourselves; (4) the ongoing dehiscence of the wound that we have called 'nature', an experience in which we continually play out separation and suture via imagined acts of violence, destruction, and attack on the one hand and nurture, growth, and merger on the other. As an aesthetic with these defining features, the eco-weird uses representational and affective strategies that counter mainstream environmental narratives. As Nicole Seymour has suggested, mainstream environmentalism tends to communicate conservative, heterosexist, racist, and moralistic imperatives, and these have been ineffective in prompting action on climate change (Seymour 2018). Indeed, mainstream environmentalism more often provokes feelings of guilt, which turn quickly to defensiveness and accusations of hypocrisy. The eco-weird is an alternative and potentially transformative aesthetic mode in that it does not revolve around moralistic and normative imperatives but exposes and explores ecological experience.

Using this definition of the eco-weird, I propose that games are weird and that they are a way of 'being ecological' (Morton 2018), by which I simply mean participating in the ecological mesh of relations that is 'the environment'. As a weird aesthetic, and, therefore, an affective experience, gameplay draws attention to our being ecological, to human enmeshment in a fragmented yet tangled mess of relations between critters and objects of all kinds. There is something about gameplay across the tabletop that generates the sensory and affective impressions that are germane to the feelings of enmeshment prompted by the eco-weird. In making this claim, I draw on the insights from game studies' material turn and its concern with play as a more-than-human phenomenon (Germaine 2020; Sicart 2021; Germaine and Wake 2022). This work establishes a dialogue between game studies and various forms of philosophical 'weirdisms' that look slantwise at materiality, including Graham Harmon's 'weird realism' (2012), Erik Davis' 'radical empiricism' (2019), and Karen Barad's account of 'queer performativity' (2011). Examples of engagement with enmeshed and weird materiality in game studies include Paul Wake's reading of games as more-than-human 'performances', in which he describes the human player as a 'component part' brought into being through the 'interpenetrating and mutually transforming interplay of a network of other component parts' (2022, p. 60). Wake gives the evocative example of the Balinderry game board, a material object from tenth-century Europe that depicts two figures engaged in a game of Tafl (Viking Chess). The board, Wake suggests, 'sees the line between hand and board dissolve in a peculiarly intimate merging of human and nonhuman components' (2022, p. 59). In his essay on the sensory experience of tabletop roleplaying games, focussing on touch, Jack Warren also considers the more-than-human dimension of play. He describes roleplaying as a 'gyre, an endless system of circulating materials and imaginings' that engenders queer intimacies between all the bodies and objects at play (2022, p. 67). My contention that gameplay is a more-than-human material phenomenon, an entanglement of agencies in a continual process of becoming, likewise elicits weird propositions that run counter to humanistic accounts of play. In my account of live-action roleplaying games (LARP), for example, I describe the occult rituals enacted in games as 'technologies' that 'puncture and disrupt the domain of ordinary reality' (Germaine 2022b). In this description, I take equally seriously the fictions, immaterial elements, and material props used in LARP, arguing that gameplay becomes a mode of weird 'magic' that reveals contamination and consubstantiality between the animate and inanimate, human and nonhuman, and between the 'inside' and 'outside' of human experience (Germaine 2022b).

As these accounts from material game studies suggest, games are not only texts and media products but intimate experiences. They provoke what Gordon Calleja calls 'involvement', a mode of engagement or attention that comprises a 'multiplicity of experiential dimensions' (2022, p. 29). These experiential dimensions are not only cognitive (requiring apprehension of game rules and narrative) but include emotional reactions, haptic, and sensory responses to the visual and tactile elements of the game, social interactions generated around the table, and more. Tabletop games are multi-layered experiences that can reveal more readily than other media, such as novels or films, that the weird is not only a mode of cultural production but, as Davis suggests, a mode of 'affect and possibility' (2019, p. 30). Moreover, accounts of material game studies reveal that the weirdness of games emerges precisely because of their discontinuous but interpenetrating experiential dimensions: material, spatial, temporal, and imaginative. As suggested above, through my description of the many (real and imagined) bits and bobs involved in games, play across the tabletop entails multiple and shifting boundary sites that temporarily structure intimate networks of action: the table, the game board, the GM's instructions, bodily and verbal gestures of other players, social frames, the rules. Across and between these borders move a multitude of material and imagined actors, producing the game as a continually shifting performance within which, as in weird experiences, distinctions such as that of subject/object break down. As the case study in this chapter will demonstrate, human players of games, as in other weird experiences, must renegotiate their positionality, their dwelling, as the game performance shifts around and through them.

9.3.1 Making a Raid on the Real

In the case study, I claim that games are not only media, or tools, for translating the virtual; they are weirder than that. They exert their own agency and presence in the world. This takes further a critique of Juul's proposition that games are 'half' real. Games are not only real; they affect the reality from which they are supposedly bracketed by the 'magic circle' of play, creating warping and torsion effects. Warren's account of roleplaying gestures to this when he emphasises the 'leaky' character of roleplay, exploring how it 'spills' its affective gestures out into the 'real world' (2022, p. 73). In his account of roleplaying games, Nick Mizer goes further, arguing that role-players work together with the worlds they create to 'kidnap' reality. It is not just that the act of roleplaying creates a new reality that comes to have its own ontology, but, rather, the new realities of the gameworld project outward at the same time as players are 'abducted' into them (Mizer 2019, p. 94). Here Mizer's account of roleplaying as 'kidnapping' reality echoes Erik Davis' assertion that the weird 'makes a raid on the real', and counteracts the idea that gaming is a naïve form of escapism (Davis 2019, p. 30). Mizer describes the disruption of a demarcation between the inside and outside, arguing that a raid occurs from within the game to the world without. This is another key feature of the weird. In Mark Fisher's work, for example, weird affects emerge from the ways in which fictions denaturalise experiences of the world by exposing our 'openness to the outside', by which he means the externality of human phenomenological perception or experience as correlated by human subjectivity (2016, p. 29). Such experiences of the 'outside' prompt the uncanny realisation that 'there is no inside… I am an other, and I always was' (Fisher 2016, p. 12). In the case of games, this is the realisation that the player is not separate from the game, as in Wake's analysis of the human player as component part.

This ontological disruption of borders is a disruption of a settled sense of ecological belonging, of dwelling securely 'inside' something, which provides the feeling of being at home in the world, the security of one's habitation. In analogue games of all kinds, from tabletop board games to LARP, players are invited to enact a disruptive transformation of the rules of reality. Although games are said to be demarcated by a 'magic circle' (Huizinga 2016), a term used in game studies to describe, variously, the psychological bubble created during gameplay, the social contracts that frame the action of gameplay, and the demarcation of the sites or space of gameplay (Stenros 2012), the effects of this disruption often spill beyond the borders marked at the beginning of play. In this sense, gameplay is more akin to the workings typically associated with magic circles in ritual, simultaneously creating and collapsing physical, social, and metaphysical borders. In so doing, gameplay 'undermines human-centred accounts of play as an activity that imposes meaning on the world and instead reveals the ways in which play is responsive to, and emerges from, a world outside, or beyond, the human' (Germaine 2022b).

The kidnapping of reality afforded by games is the taking hostage of a portion of time and space, but also, at the same time, a form of hosting and of being hosted.[1] These paradoxical dimensions of games, of playing and being played, introduce the notion of the game itself as a subject, an agency, that exerts its own effects on the world, and over which players do not have control. Here, I turn to Hans Georg Gadamer's account of play, in which games are described as a mode of 'reality that surpasses' the player (Gadamer 1994, p. 109). Gadamer describes play as an impersonal agency, anticipating writing on the weird that often appeals to a similar concept of a nonhuman agency (see, e.g. Fisher 2016). Gadamer argues that all play is, really, 'being-played' (1994, p. 106). All play is movement backward and forward, but 'it makes no difference who or what performs this movement [...] it is the game that is played—it is irrelevant whether there is a subject who plays it' (Gadamer 1994, p. 104). Play is not something a person does; there is not an activity called 'play', there is only play itself. This description of play as an impersonal agency diverges from discussions in game studies of 'player agency', in which human player experience is often prioritised in the analysis of games. Rather, in Gadamer's account, attention is paid to the (weird) ways in which play unfolds independently and to its enfolding of players. The 'primacy' of the game over its players poses what Gadamer calls 'risk', which I suggest is like that which plays out in eco-weird fictions of merger (e.g. *The Willows* by Algernon Blackwood [1907] or *The Voice in the Night* by William Hope Hodgson [1907]). In the case of play, the risk entails succumbing to the game, being drawn into play, held in its spell, and kept there (Gadamer 1994, p. 106). Again, I draw attention to synergies between the weirdness of play and theories of the weird, both of which focus on 'ineluctability' as

[1] I am aware that in this evocation of the dichotomy hostage/host I raise the spectre of Derrida. Readers will either be delighted or disappointed to learn, however, that I won't be drawn in.

a key experience and on the surrender of human subjectivity to something other than itself, typically in a moment of ontological shock such as China Miéville's account of the weird as genre (2009, p. 512).

9.4 Gaming's 'Ecological' Turn

The context for this investigation of games as eco-weird experiences is an increased prevalence in ecological and weird themes in gaming. My claim is not only that games are weird, but that they contribute something particular to our understanding and experience of the eco-weird. In part, I am inspired by the burgeoning field of 'ecogames', which has seen scholars, designers, and the game industries turn their attention to the climate crisis. In addition to an increased number of tabletop games designed around the theme of sustainability, climate action, and the 'natural' world, there is a wealth of research on the possible affordances of ecologically conscious games, as well as around the risk or damage posed by games that continue to peddle agroscientific, capitalistic, mechanistic, and otherwise damaging patterns of thought and behaviour. This research has identified the concept of 'ecogames' as a way of better understanding the risks and benefits presented by games as cultural objects and social actors (Op de Beke et al. 2024). Game studies' ecological turn includes analyses of environmental game design (Raessens 2010; Chang 2019), the proposal of frameworks for the eco-ethical evaluation of games (Backe 2017; Germaine 2022a), discussions of the potential for environmental game design to influence behaviour (de Salas et al. 2022), and studies quantifying observed outcomes resulting from playing environmental games (Meya and Eisenack 2018; Fjaellingsdal and Klöckner 2019).

There is a wider application of this chapter and its claim concerning the weirdness of games, then, that contributes to this growing body of research on ecogames: what is the implication of the insight that games just *are* weird for these attempts to better understand games, or to design better games, in the context of climate change and the need for personal, social, and cultural action? In asking this question, to which this chapter can provide only a partial answer, I want to draw attention to a possible exchange between game studies and other fields of enquiry. Game studies (which is a way of paying attention to the design and play of games) can contribute much to broader critical debates about the affectivity of climate change, but it is also clear that game studies (and game designers) should draw from the insights of emerging critical fields, such as the study of the eco-weird. Working at this intersection, I identify an emerging trend in tabletop gaming towards eco-weird aesthetics, that is towards eco-weird experiences, in games. These might be generated through rules, mechanics, and components as easily as through 'weird' theming or tentacular story tropes.

9.5 How to Pay Attention: A Methodology for Game Analysis

In writing this chapter, I have adopted a particular academic style that situates 'me' outside what I describe, creating the impression of a researcher (subject) examining a game (object). However, the eco-weird is precisely an experience that makes it clear that no such external vantage point exists and that games are not objects in this way. As player, I am tangled up in, and my subjectivity is constructed through, the performance of the game. The same is true of the researcher, who is an agency of observation entangled within the phenomena they observe (Barad 2007). Reflecting upon gameplay experiences for the purpose of academic analysis, paying attention in such a way as to give a responsible account of entanglement, requires jettisoning the pretence of analytic objectivity. In developing a method for this in game studies, I am inspired by Seth Giddings' approach to game analysis, which he calls a 'microethology' (2009). 'Ethology', following Deleuze, eschews the anthropocentrism of ethnography and offers instead a description of the multiplicity of participants who play. Although I recognise that accounts of games are filtered through the human phenomenological experience, and, indeed, that the eco-weird names precisely this phenomenological experience, ethology is the study of relations beyond the human, of the capacities for affecting and being affected that cut across bodies, objects, machines, and agents of all kinds, a study of mutual becomings (Deleuze 1992, pp. 627–8; Giddings 2009, p. 152). A microethology of gaming identifies and attends to the relationships between technology, materials, bodies, play practices, virtual worlds, imagined characters, and more, taking account of the 'collusions' that generate play experiences (Giddings 2009, p. 156). Games are no longer 'objects' of research. Indeed, secure distinctions between people, cultural objects, technologies, and materials are undone if the researcher pays attention to the slippages that occur between subjects, identifications, and perspectival positions (Giddings 2009, p. 155). As Giddings suggests, such work requires a non-scientific, improvised, and opportunistic approach to attending to gameplay experiences.

Another aspect of my methodology comes from descriptive phenomenology. Simply put, this requires asking what an experience is like, attempting to put aside theoretical frames and assumptions, and instead attending to the perceptions, thoughts, and emotions that arise during the experience. Although I have approached this study from the outset with the eco-weird in mind, and I make no pretence to bracket out my thinking on that as I play, in my game sessions I am paying attention to what it feels like, attending to the actions and movements of my body, and those of others, to what I think about, whether it is frustrating, boring, tiring, exciting, uncomfortable, and so on. Paying attention to emotional and embodied experience in this way is increasingly of interest to researchers of the weird in cultural and literary studies, in part because textual analysis alone cannot capture the affective dimension of the works. This attention to *experience* appears, for example, in Kaisa Kortekaillio's work on ecological science fiction, which focuses on the experience of reading what she calls 'mutant narratives'. Kortekaillio develops a 'performative enactivist' approach that draws on autoethnographical explorations of experience and that takes the 'experiential self' that is reading, not the text, as the locus of research (2020, p. 34). For Kortekaillio, reading is a bodily practice, and attention should be paid to its material and more-than-human dimensions (2020, p. 19). Moreover, paying attention to the bodily practice of reading (or, as is the case here, playing) means understanding that environments are not stable but exist in continuous processes of change and interchange (Kortekaillio 2020, p. 20).

As this eco-weird project focuses on games, which are more-than-human performances, the human-centred focus of descriptive phenomenology must be tempered with an awareness of the more-than-human nature of perception; it must consider the other agencies and experiences that emerge in gameplay. As I have claimed, games involve much more than just a human experience, and the case study included here aims to explore these more-than-human dimensions of play, paying attention to the participation of nonhuman agencies (rules and mechanics), playing pieces, and space. The case study is, therefore, a result of the application of more-than-human phenomenological description, which requires paying attention to the multiple actors involved in perceptions of gameplay and a microethology of play that attends to the physical bits and bobs and to the technologies (or, mechanics) of the games. I recorded my reflections on gameplay in an unstructured way following gaming sessions, noting ideas, feelings, and observations as they occurred. Further reflections were recorded in an opportunistic way, during conversations with other players that occurred after the fact, in which we discussed our playthroughs of the games. I constructed the structured case study after establishing key themes and insights from these rough recordings, a process that was undertaken in dialogue with the critical and theoretical material on game studies, the eco-weird, and aesthetics discussed above.

9.6 Sleeping Gods

This case study investigates the weird experience of 'immersion' in story games, which are a genre of big box games that use boards, miniatures, cards, and other components. They typically combine eurogame mechanics (resource management, action economy, worker placement) with features from combat games (die rolling, miniatures, maps) and roleplaying. Story games draw on influences from the dark fantastic in literature and film. The publisher Fantasy Flight, for example, boasts a backlist featuring IP such as *Lord of the Rings* (fantasy), *Battlestar Galactica* (science fiction), and *Arkham Horror* (the weird). However, rather than focussing on examples such as *Mansions of Madness*, which adapt Lovecraftian fiction and are obviously 'weird', I identify how story games immerse players in *enweirded* environments beyond recourse to thematic content and literary references. For this purpose, I examine gameplay sessions of *Sleeping Gods* (2021), a game designed by Ryan Laukat.

Sleeping Gods boasts a large gameworld that is encountered via a mapbook, which functions as a game board. Each page has dozens of locations. At each one, players encounter multiple branching storylines. The story opens with players taking control of a cast of characters who, having begun a sea voyage to New York, find their ship afloat on a strange ocean amid an archipelago, somewhere in another world. This is a secondary world following the tradition of high fantasy, and it is reached via a portal of some kind—though the mechanics of transportation from one world to another are not revealed. There is a nod to 'Lovecraftian' themes when player characters learn of the 'sleeping gods' with whom they must somehow communicate if they wish to return home. The lusory goal of the game remains dependent on the story nodes that players encounter. In our playthrough, the inhabitants of the archipelago warned of the potentially disastrous consequences of waking the gods, instituting a moral quandary and an ethical choice into the unfolding of the endgame. The genre is high fantasy. It contains combat encounters with sea monsters, artefacts and totems that must be collected, and multiple questlines to discover. The 'open' world feeling of the game is generated by the 26-page map-book, each page containing dozens of numbered locations, corresponding with 250+ entries in a storybook. The 250+ nodal points form a complex branching narrative, with each entry containing multiple options depending on what objects and keywords you have found so far. In this mechanic, *Sleeping Gods* draws on the *Fighting Fantasy* and *Choose Your Own Adventure* gamebooks popular in the 1980s and 1990s, though here the navigation of numbered entries in a branching story is enacted spatially, via the map, across which a ship miniature moves, and player choice is constrained within movement rules (which locations can be reached) that rely on 'craft' tests that require you to deplete your characters' actions to use their craft skill and then add this skill to a number drawn randomly from the card deck. In addition to fighting monsters and making decisions about what to do at each location, players collect 'quest' cards containing keywords that build up a compendium, allowing them to unlock options for different storylines at other locations. They also collect artefacts (represented on cards) that give the characters 'buffs' for making tests and surviving combat. Card draws make the outcomes of tests feel randomised, with some control available to players in which characters they choose to enter the tests and where to assign buffs. Taken together, these disparate elements (the branching narratives of the storybook, map navigation, keyword collection, item collection, and card-draws) create an emergent narrative unique to each playthrough of a game. The game has many components, including different card decks, tokens, and multiple boards, which take up a large amount of table space and require significant cognitive attention in their continual management.

Play sessions of *Sleeping Gods* are long, especially if you want to reach one of thirteen possible endings. Our game comprised 36 rounds played over three days in multi-hour-long (typically three or four) sessions. Between these sessions, we left the game on the kitchen table. There was the option to 'save' the game and re-box the components, which we opted not to take. Immediately, we felt that setting up the game, putting it away, and setting it up again was enough of a rigmarole that, if we were committed to seeing the story through to the end, we better not put it away at all. The result was that the game disrupted our lives for several days. There is only one table big enough in my house, which meant that we had to eat meals at the coffee table and edge carefully around the kitchen table so as not to disturb the set-up. We also felt guilty for not getting back to the game when we elected to do something else, like watch TV. We talked about how the game needed us to finish it, and that in committing to the playthrough we were now trapped with the game, and its disruptive presence in our home until it was done. The length of each play session exceeded what we expected, and we would play for three hours at a time wondering if we were yet near the end. This is not to take away from the enjoyment we found in the story and the setting but to emphasise that, despite the appeal of the game, the demands it made on time and space, along with the ineluctable quality of having to play until the end, were very much part of the experience and that the 'kidnapping' of reality achieved by the game was not only the result of our absorption in its storyworld but also the coercive demands it made on our time, bodies, and domestic space.

9.6.1 *Submersion/Immersion/Surfacing*

A recurring theme of reflection during and after play of *Sleeping Gods* was immersion: it was taking up a lot of space in our lives, but did we really feel part of the world of the game? As game scholars, we were struck by how the ocean setting of the game aligns with the 'watery' metaphors used in game studies to explain how games involve players in their storyworld. In the recent discussion of video game versus board game immersion cited above, the writers use the word 'submergence' to describe the feeling of being fully involved in a narrative setting (Farkas et al. 2020). For 'submergence' to occur, they argue, the mediating technologies require fidelity and hyper-realism. However, this idea of immersion ignores a key element of the process, famously described by Janet Murray (2016 [1997]). Murray's theory of immersion derives from the physical feeling of being plunged into an ocean or a pool. Crucially, in this metaphor, the awareness of passing through a surface and into the water, and then of being surrounded by water, are paradoxically necessary for immersion. As Wake contends, such an idea of immersion relies on player awareness of a threshold object (e.g. the screen) that separates the world of the player from that of the game. In tabletop games, these threshold objects (tokens, boards, dice, and cards) are precisely what 'draw the player into the gameworld and maintain its boundaries as a distinct space' (Wake 2019). Calleja also emphasises this need for active awareness of immersion in what he describes as the player's 'transportation' from the 'real' world to the game world (2022, p. 201). 'Transportation' is the act of sharing fiction and 'requires a negotiation of, and agreement on, what is going on in that fictional world, and how the actions taken in it will change the state of that fictional world, not just for the actor but also for the others inhabiting it' (Calleja 2022, p. 202). In *Sleeping Gods*, our characters were themselves 'transported' to another world, and we, in turn, as players, transported the plastic ship across the board. Our experience of transportation was therefore doubled and overt: as our characters, we told the story that unfolded in the world of *Sleeping Gods*, and, as players, made strategic decisions about how best to navigate the archipelago. As Wake (2019) notes, 'to be immersed in analog storyworlds is [...] to occupy two spaces simultaneously, and at points it becomes difficult to distinguish between "inside" and "outside"—the game and the metagame'. Certainly, in *Sleeping Gods*, transportation prompted this dual awareness: the feeling of being in a storyworld and in a game. Where Wake emphasises the bits and bobs that produce immersion, Calleja (2022, p. 202) identifies the 'mechanical system [...] as a mediator of this negotiation', arguing that it serves 'as a form of reality engine'. In both accounts of immersion, as in our experience of playing *Sleeping Gods*, the emphasis is on the movement of play, those initiated by the player, but also those demanded by the game itself, by the bits and bobs and by the 'engine' of the rules, all of which propel the ship across the map, taking us to places we didn't necessarily choose to go. Our discussions of immersion in the gameworld, then, quickly came to focus on the negotiations of play, on the relations out of which the reality of the game emerges, and we had to take into account all the other 'players' involved, including the game itself.

Discussions of whether we were immersed in *Sleeping Gods* revealed that 'submergence' in its oceanic world involved 'surfacing' as much as 'plunging', a push and pull process that continued throughout the play sessions. It is in this negotiation between immersion and surfacing that an eco-weird experience emerges. As Wake (2019) suggests, the paradoxical nature of immersion requires a continual engagement with the 'mess of the tabletop', and is, in fact, a result of 'the tokens of analogue games, which ostensibly anchor the player within the horizon of the gameworld, [whilst] simultaneously gestur[ing] (vertically) to the world of the player'. Wake (2019) describes this aspect of board game play as 'unsettling' yet productive, arguing that 'play takes place in an unsettling, because unsettled, third game space that emerges when play oscillates between the tabletop game space and the space occupied by those playing the game'. With *Sleeping Gods*, there is a great deal of 'mess' on the table, as well as complex mechanics that continually prompt awareness of the 'oscillation' that Wake describes. Furthermore, thinking through game immersion with the eco-weird in mind made us consider that there were not two, nor even three, distinct 'worlds' between which we travelled as players, but, as the game sat stubbornly on the kitchen table for days, something more like a mesh of worlds that became increasingly tangled. The game was a messy 'threshold', the interface of which continually shifted position as game pieces were touched, tokens drifted from their position on the board, drinks were spilled, and cards knocked onto the floor. Just as a player's inhabitation in the gameworld is unsettled by the tactile materials and mechanics of board games, facilitating immersion in the game as experience, so does the game as experience unsettle a player's dwelling in the 'real' world. Both the material existence of the game, its bits and bobs, and the imaginative space it required, leaked out into the domestic space of the kitchen. If it is as Wake (2019) suggests that we understand immersion in board games not in terms of the transportation of the players to the world of the game, 'but in a broader sense in which the gameworld extends beyond the game and into a zone that is proximal with that of the player's actual world', then it is this proximity that is the locus of eco-weird experience. Our experience of immersion in *Sleeping Gods* was not a simple question of whether we felt 'transported' to its storyworld but became an issue about the bleed between, and tangle of, proximal worlds, the taking over of one world by another, and the coercive nature of gameplay. The complex mechanics of the game, multiple components, and an involving storyline worked together to erect and disrupt environmental boundaries that included those represented in the mapbook, as well as those that emerged on and around the tabletop. As virtual worlds leak into actual worlds, and dwelling becomes unsettled, the eco-weird game draws attention to our situatedness in an environmental mesh. It functions, like the eco-weird oceans in literature, as a contact zone with 'flows of substances and agencies' that are more-than-human (Germaine 2023, pp. 130–1). Thematically, *Sleeping Gods* is about being unhomed, about being transported to another world, but my reflection on play considers how this theme is at the same time manifest in the dynamic and mechanical aspects of the game, spilling out, enweirding the mundane space of the kitchen, remaking the tabletop, and unhoming the player.

9.6.2 *Loopy Causality*

As well as being a simultaneously immersive and unsettling experience, playing *Sleeping Gods* also became eco-weird because of the spatial and narrative loops in which gameplay was caught. Two related reflections emerged in discussion about these loops: the first was about the overall impression of the game as a contingent and unstable narrative that could have gone down very differently, and the second concerned the twisty, entangled nature of causality as experienced during gameplay. These eco-weird experiences occurred despite, or perhaps even because of, the mechanical and systematic arrangement of space and time as set out in the mapbook and storybook. Reading these documents as documents, rather than playing the game, presents *Sleeping Gods* as a series of causal sequences, which are numbered (both on the map and in the book) and arranged logically (*if* you have keyword 'a', *then* consult entry 'b'). However, this is not what play feels like. Rather, the play experience recalls Davis' description of the weird as 'the loopy, twisty, tricksy ways whereby things come to be' (2019, p. 30). We felt from the outset that it would be impossible to navigate with any sense of mastery the gameworld. It was overwhelming to realise that there were twenty-six pages of map, far more locations than we would be able to visit in thirty-six turns. After a few turns, we also realised that none of the locations so far discovered combined into a straightforward narrative thread. Instead, we had found a collection of discontinuous leads pointing in different compass directions. For some hours of play, no story resolution presented itself, nor was there a clear lusory goal that gave a sense of what 'winning' the game would entail. The 'tricksy' feeling persisted throughout, particularly when the ship travelled in circles. However, it was most obvious at the end of the game. When we did find one of the possible endings, we knew it was only one possible ending, but we had no sense of what the other twelve entailed. Nor did we feel that we had, at any point, made a deliberate choice to follow a path to this particular ending. Instead, narrative resolution was the emergent result of the loopy route the ship had taken around the map. We knew that playing again would give a different experience and that we would have a little more familiarity with the gameworld, though not nearly enough to navigate it more efficiently. Map pages took multiple turns to cross, and we covered only six pages in all, visiting a handful of locations on each, less than 10% of the total. It was clear that if we began the story again, which players are invited to do, we would encounter different routes that would generate different story arcs and, so, offer different versions of winning and losing, entailing different moral choices. The game was, therefore, a mutating narrative, shifting and contingent, with each playthrough writing over the 'original' story, despite the fact that all the mechanical elements (locations, monster encounters, artefacts, etc.) would remain the same. Playing again would not generate cumulative experience, or 'XP', to use the word in its gaming sense, but entirely new aesthetic experiences requiring different ethical orientations.

The second reflection about causality in *Sleeping Gods* concerned how narrative and spatial sequences felt during play. As discussed, the number of locations, their spatial arrangement, and the impossibility of their efficient navigation resulted in our ship travelling in circles, in debates about where we'd been, what compass direction to follow, as well as feelings of repetition and déja vu. The latter emerged both because of the duration of gameplay (have we been here before?) and because of mechanical constraints (entering a location where you don't have the right keyword to unlock the story can trigger a similar combat encounter to another location because the game is 'filling' gaps for players who haven't collected the right clues). Despite knowing that the game was, ultimately, arranged logically, the feeling was of frustrating complexity. The keywords themselves are cryptic, giving the impression of unknown stories playing out in the background of the game, of machinations that would affect our journey in unexpected ways. In giving rise to feelings of disorientation and a lack of agency, of glimpsing story threads we could not follow, the gameworld became like a dreamworld, rather than a fully drawn secondary world. Mechanically, we encounter the story in fragmented and disjointed narrative units, via keywords that are not always directly meaningful, but rather impressionistic, and by visiting locations 'out of order'. All this added to the overall sense of an *enweirded* causality rather than a mechanistic one. In addition to considering the loopy, repetitive nature of play in terms of story and gameworld causality, we wondered about the experience of play should we take up the invitation to commence a second playthrough of the game. Although we would have collaborated with the game to produce a different narrative, a different version of the world, would the feeling of play have been the same? Perhaps the invitation to play again, to explore an alternative narrative thread was, really, a ruse on the part of *Sleeping Gods* to kidnap us a second time. Once we had been dragged back into its world for a second run, we might find that our experience of completing the mechanical tasks of game management, of acting on behalf of the game environment against the characters in the story, of getting lost in the map, would create an even stronger sense of *déjà vu* than in the first game, all of which enhanced our feeling that the game had us caught in a twisted loop of time and space. Perhaps we better quit while we could…

As an eco-weird experience of loopy, tangled, and non-sequential space and time, *Sleeping Gods* performs what Morton calls 'aesthetic' causality. In his theory of dark ecology, Morton suggests that 'weirdness means the turning of causality' (2016, p. 13). He identifies aesthetic causality as a way of thinking through how 'things influence one another such that they become entangled and smear together' (2016, p. 84). *Sleeping Gods* generates just such an experience of aesthetic causality. What happens is a result of what Barad describes in her work on causality as 'entangled agencies' (player choice, story prompts, game mechanics, the chance of a card draw), and gameplay comprises a series of evolving 'intra-actions' within the phenomena that is the game, rather than a series of player interactions, or choices (Barad 2007, p. 131). Cause and effect emerge through intra-actions, not through a linear sequence in which one agent, or object, acts upon another (Barad 2007, p. 193). In this account of intra-activity, causality is not a binary affair, a choice of causal determinism versus absolute freedom. Of course, *Sleeping Gods* is constructed to emphasise player choice, but the ergodicity, the need for players to provide active input into the game engine, is subsumed into complex feedback loops of intra-activity. Multiple branches emanate from a single node: some lead to the acquisition of resources, some open new storylines or quests, some point to new locations not yet visited. At the same time, the randomised card draw element of 'craft' tests determines how far the ship can travel across the map. Thus, player choice is just one 'cut' to be made within the phenomena of the game, and it is already contingent on other cuts already made, part of an evolving phenomena, and not the determining factor of how things will play out. As suggested, the cuts made within the phenomena of the game, the intra-action of player choice, card deck set-up, card draws, and mechanical constraints produce very different 'worldings' of *Sleeping Gods* in each game.

9.6.3 Whose Game? Rethinking Player Agency

The submergence of player agency into the phenomena of the game reveals something crucial about 'agency': it cannot be held by an individual but is instantiated in actions and relations, in shared becoming. In terms of story generation and following narrative threads to achieve a 'win' condition, the CYOA (choose your own adventure) element of *Sleeping Gods* would seem to favour something game studies has called player agency. However, our playthrough revealed player agency to be a fiction. Unlike a CYOA novel, in which player choice (if not agency) determines the shape of the story, in *Sleeping Gods* players submit to the reality engine of the game. Traversing the map with a little plastic ship by undertaking craft challenges is a cumbersome way of turning pages and, moreover, constrains player choice around where to go next (a choice that initially seems very open compared to a CYOA novel) very quickly. The initial 'player versus environment' framing of the game is complicated through questlines that cast ethical doubt over players' intentions to wake the gods. Moreover, these ethical choices, and their framing, are determined by the locations that players can visit. Given the information we had found on our journey, we felt we were on the 'right' path, in terms of the moral framing of our characters' choices and in terms of how to win the game. Actually, though, we were making 'choices' according to the demands of a particular storyline onto which we had only stumbled due to the way game mechanics had determined our movement. Increasingly, then, the initial 'player versus environment' framing gave way, and we felt our actions as players were subordinate to the agency of the game itself. This feeling of being controlled by the game itself, rather than being in a contest with the game environment, suggested, as I describe above with reference to Gadamer, an impersonal agency within which all elements of the game were enfolded. As Gadamer suggests, 'the players are not the subject of play; instead play merely reaches presentation (*Darstellung*) through the players' (1994, p. 103). This feeling was apparent from the introductory encounter, a walkthrough that is designed to help players learn the rules of the game. None of the individual game mechanics were unfamiliar to us, so the walkthrough should have been straightforward—a matter of figuring out the combination required for this game. However, it felt very weird. We followed the rulebook's instructions to set up the game and lay out the components, and we took the first few turns following its directives, without any sense of why we were moving this token there, taking this test, or drawing that card. Nothing we did seemed to have any meaning but felt instead automota-like. In terms of the gameworld, we did not know where we were going on the map, nor what the overall lusory goal was supposed to be. Learning the game was at once a highly mechanical and disorienting experience, and we felt like cogs in the working of the game, rather than feeling like we might become masters of its mechanics and strategy. There is a definite sense of being controlled by the game of *Sleeping Gods*, then, that mirrors the thematic of being lost at sea in an unfamiliar world. Not only was this a game of player versus environment, but also a game in which players mechanically carry out the whims of the environment against themselves, drawing disaster cards from an events deck, applying ship damage tokens to their own ship, and failing to navigate across the treacherous sea in one turn. Here the game environment is not possessable nor passive but recalls Gadamer's suggestions that 'the playing field on which the game is played is [...] set by the nature of the game itself and is defined far more by the structure that determines the movement of the game from within than by what it comes up against' (1994, p. 107).

9.7 Conclusion: Let's Play (Again)

The case study in this chapter examines one game performance, taking a genre of tabletop games as its example, the story game. Broadening out from the specific case study, here I suggest conclusions about games as eco-weird experiences. Tabletop games kidnap reality, taking over space, filling time, and coercing human players to enact their demands and work their mechanics. The material components, game rules, and imagined part(icipant)s spill across the tabletop, leak into, and contort the reality around themselves, discomfiting and disrupting intimate and familiar sites of dwelling. Thinking of games in this way, as disruptive presences that demand the surrender of human agency, sits in tension with fantasies of mastery (over terrain, economic power, defeated enemies, and so on) that games often promise. Games are not engaging because they allow human players to enact fantasies of mastery but, rather, as Gadamer suggests, 'the attraction of a game, the fascination it exerts, consists precisely in the fact that the game masters the players' (1994, p. 106). Gadamer also defines play not as an activity or an attitude, something that takes place in a player's consciousness, but as a process that occurs 'inbetween'. It is this inbetweeness, I argue, that is the province of the eco-weird. Being transported into, or becoming immersed within, a game requires negotiating multiple borders, material and imaginative, and asks players to cultivate an awareness of moving on behalf of, and among, others, whether this be other humans, playing pieces and tokens, or imagined beings. Gameplay thus demands the constant renegotiation of the player's positionality and a resituating of oneself within the game and in one's dwelling, as inside turns to the outside, and back again, and submergence meets surfacing. Games promote constant contact with surfaces, boundaries, and edges, contact that is, paradoxically, required to facilitate game immersion. All of this serves as a reminder that participation in worlding is a relational affair and that reality is neither smooth nor a bounded space nor loyal to human perceptions but a difficult matter of ongoing relationality and making. Gaming is co-worlding, an invitation (imperative?) to participate in making new worlds. In *Sleeping Gods*, what emerged from the process of play was not a definitive story, nor a sense of having won the game, but a weird sense that the world we had made together was waiting to be re-made, if only we would play again.

The weirdness of gameplay will generate different experiences in different situations, but it is always one that confronts human players with the strangeness of their subject position in relation to the more-than-human others with whom they play and of their (in)habitation of a gameworld and, sometimes, the actual world. These are also experiences that, as players handle game tokens, draw cards, place map tiles, or thumb through difficult rule books, offer an experience of co-worlding with a host of participants, some of whom are present in the room, and others who reside in the virtual, exerting their own strange agency through the performance of the game. Gaming is weird, then. It is also ecological. Being ecological is realising there are other ways of living, that other patterns of thought and habitation are possible, that we are in fact living somewhere rather than nowhere, and that we make that world just as it makes us. I have argued that gaming offers a space of reflection on this weird situation and can generate a possibility space for confronting and renegotiating the boundaries and points of connection that comprise ecological reality. The eco-weird is an encounter, a realisation that being is being within an ecological mesh. As a mode of weird worlding, gameplay is fluid and highly contingent. Worlding unfolds differently each time we play, and this is because of the interaction of rules and constraints, not despite them. The weird encounter is in part about the meeting of deterministic (rules) with probabilistic mechanics (such as dice) and free roleplay.

There are ways, however, in which games push back against, or might work to contain, the eco-weird. This is most obviously apparent in the escapist, self-congratulatory, or colonial themes offered by many tabletop games, but we also encounter it in the social environments that develop around gaming, in the exclusive and discriminatory behaviours that can sometimes fester there. My claim that games are eco-weird experiences, then, should not be taken as a wholehearted endorsement of gaming as a radical or emancipatory practice that will lead, in itself, to a transformation of self, society, and the environment. Just as Kortekaillio notes that her 'mutant narratives' will not 'give rise to new kinds of posthuman subjects or open up new posthuman worlds of meaning', nor will gaming itself bring about an ecological revolution (2020, p. 17). However, I will claim, just as Kortekaillio does in relation to reading, that games might help human players to 'experience the world in richer and more nuanced ways' and encourage them to do the difficult work of what Haraway describes as 'staying with the trouble' of ecological belonging' (Kortekaillio 2020, p. 17; Haraway 2016). However, games will only do this work if we pay attention to how we play, giving reflective and analytical space to the eco-weird experiences that can arise across the tabletop, to moments of discomfort and strangeness. By paying attention in these ways, those interested in cultivating ecological thought and subjectivity through play can draw attention to how games reveal human enmeshment in more-than-human worlds.

Acknowledgements I am grateful to the organisers and attendees of the inaugural symposium for the Society for the Study of the Eco-Weird for their comments on a draft of this paper and for suggestions that improved its argument. I am also grateful to Paul Wake. His work in game studies has not only been invaluable in the development of the concepts in this chapter, but he also agreed to play *Sleeping Gods* with me and to talk about it a lot! The ideas presented here are as much his as mine.

References

Backe, H. 2017. Within the Mainstream. *Ecozon@* 8 (2): 39–55.

Barad, K. 2007. *Meeting the Universe Halfway: Quantum Physics and the Entanglement of Matter and Meaning*. Durham, NC: Duke University Press.

———. 2011. Nature's Queer Performativity. *Qui Parle* 19 (2): 121–158.

Calleja, G. 2022. *Unboxed: Board Game Experience and Design*. Cambridge, MA: The MIT Press.

Chang, A. 2019. *Playing Nature*. Minneapolis, MN: University of Minnesota Press.

Davis, E. 2019. *High Weirdness: Drugs, Esoterica, and Visionary Experience in the Seventies*. Cambridge, MA: The MIT Press.

Deleuze, G. 1992. Ethology: Spinoza and Us. In *Zone 6: Incorporations*, ed. J. Crary and S. Kwinter, 625–633. New York: Zone Books.

———. 2014. *Difference and Repetition*. Translated by P. Patton. London: Bloomsbury.

Ermi, L., and F. Mäyra. 2005. Fundamental Components of the Gameplay Experience: Analysing Immersion. In *Changing Views: Worlds in Play. Selected Papers of the 2005 Digital Games Research Association International Conference*, ed. S. de Castell and J. Jenson, 15–27.

Farkas, T., S. Wiseman, P.A. Cairns, and R. Fiebrink. 2020. A Grounded Analysis of Player-Described Board Game Immersion. *Proceedings of ACM CHI Play* 2020: 427–437.

Fisher, M. 2016. *The Weird and the Eerie*. London: Repeater Books.

Fjaellingsdal, K., and C. Klöckner. 2019. Gaming Green. *Frontiers in Psychology* 10.

Gadamer, H. G. 1994. *Truth and Method. Second, Revised Edition*. Translated by J. Weinsheimer, and D.G. Marshall. New York: Continuum.

Germaine, C. 2020. Encountering Weird Objects. In *Diseases of the Head*, ed. M. Rosen, 361–394. Punctum Books.

———. 2022a. Nature Games in a Time of Climate Crisis. In *Material Game Studies: A Philosophy of Analogue Play*, ed. C. Germaine and P. Wake, 143–164. London: Bloomsbury.

———. 2022b. The Magic Circle as Occult Technology. *Analog Game Studies* 9(4), https://analoggamestudies.org/2022/12/the-magic-circle-as-occult-technology/. Accessed 7 May 2024.

———. 2023. *The Dark Matter of Children's Fantastika Fiction: Speculative Entanglements*. London: Bloomsbury.

Germaine, C., and P. Wake. 2022. *Material Game Studies: A Philosophy of Analogue Play*. London: Bloomsbury. Edited volume.

Giddings, S. 2009. Events and Collusions: A Glossary for the Microethnography of Video Game Play. *Games and Culture* 4 (2): 144–157.

Grosz, E. 2004. *The Nick of Time: Politics, Evolution and the Untimely*. Durham, NC: Duke University Press.

Haraway, D. 2016. *Staying with the Trouble. Making Kin in the Chthulhucene*. Durham, NC: Duke University Press.

Harman, G. 2012. *Weird Realism: Lovecraft and Philosophy*. Winchester: Zero Books.

Huizinga, J. 2016 [1949]. *Homo Ludens: A Study of the Play Element in Culture*. Kettering, OH: Angelico Press.

Juul, J. 2005. *Half-Real: Video Games between Real Rules and Fictional Worlds*. Cambridge, MA: The MIT Press.

Kortekaillio, K. 2020. Reading Mutant Narratives: The Bodily Experientiality of Contemporary Ecological Science Fiction [PhD Thesis]. Helsinki: Helsingin Yliopisto.

Merleau-Ponty, M. 2007 [1958]. Labyrinth of Ontology. In *The Merleau-Ponty Reader*, eds. L. Lawlor and T. Toadvine, 415–418. Evanston, IL: Northwestern University Press.

Meya, J., and K. Eisenack. 2018. Effectiveness of Gaming for Communicating and Teaching Climate Change. *Climatic Change* 149: 319–333.

Miéville, C. 2009. Weird Fiction. In *The Routledge Companion to Science Fiction*, ed. M. Bould, A. Butler, A. Roberts, and S. Vint, 510–517. London: Routledge.

Mizer, N. 2019. *Tabletop Roleplaying Games and the Experience of Imagined Worlds*. London: Palgrave Macmillan.

Morton, T. 2016. *Dark Ecology: For a Logic of Future Coexistence*. New York: Columbia University Press.

———. 2017. *Humankind. Solidarity with Nonhuman People*. London: Verso.

———. 2018. *Being Ecological*. London: Penguin Random House.

Murray, J. 2016. *Hamlet on the Holodeck: The Future of Narrative in Cyberspace. Updated Edition*. Cambridge, MA: The MIT Press.

Op de Beke, L., Raessens, J., Werning, S., and G. Farca. 2024. *Ecogames: Playful Perspectives on the Climate Crisis*. Amsterdam University Press. Edited volume.

Raessens, J. 2010. Ecogames: Playing to Save the Planet. In *Cultural Sustainability*, ed. T. Meireis and G. Rippl, 232–246. London: Routledge.

de Salas, K., et al. 2022. Improving Environmental Outcomes with Games. *Simulation and Gaming* 53 (5): 470–512.

Seymour, N. 2018. *Bad Environmentalism: Irony and Irreverence in the Ecological Age*. Minneapolis, MN: The University of Minnesota Press.

Sicart, M. 2021. Playthings. *Games and Culture* 17 (1): 140–155.

Stenros, J. 2012. In Defence of a Magic Circle. *Proceedings of a DiGRA Nordic 2012 Conference: Local and Global—Games in Culture and Society*. http://www.digra.org/wp-content/uploads/digital-library/12168.43543.pdf. Accessed 7 May 2024.

Toadvine, T. 2012. Enjoyment and Its Discontents. On Separation from Nature in Levinas. In *Facing Nature. Levinas and Environmental Thought*, ed. W. Edelglass, J. Hatley, and C. Diehm, 161–189. Pittsburgh, PA: Duquesne University Press.

Wake, P. 2019. Token Gestures: Towards a Theory of Immersion in Analog Games. *Analog Game Studies* XI (1). https://analoggamestudies.org/2019/09/token-gestures-towards-a-theory-of-immersion-in-analog-games/. Accessed 7 May 2024.

———. 2022. Component Parts: Board Games as Architecture and Performance. In *Material Game Studies: A Philosophy of Analogue Play*, ed. C. Germaine and P. Wake, 47–64. London: Bloomsbury.

Warren, J. 2022. A Queer Touch of Fantasy Role-play. In *In Material Game Studies: A Philosophy of Analogue Play*, ed. C. Germaine and P. Wake, 65–82. London: Bloomsbury.

Ludography

Blood Bowl. 1986. *Designer: Jervis Johnson.* Games Workshop: Publisher.

Forests of Pangaia. 2023. *Designer: Thomas Franken. Artists: Marcel Domke, Chris Karbach, Anna Meshchanova.* Publisher: Pangaia Games via Kickstarter.

Sleeping Gods. 2021. *Designer and Artist: Ryan Laukat.* Red Raven Games: Publisher.

CHAPTER 10

Forms and Themes of the Eco-Weird: Experimentation and Play in a Warming World

Gry Ulstein

10.1 Introduction

In *Bad Environmentalism: Irony and Irreverence in the Ecological Age* (2018), Nicole Seymour discusses an eccentric collection of artifacts, from *Wildboyz* (2003–2006; a spin-off TV series of MTV's more famous *Jackass*) to *The Simpsons Movie* (2007). Seymour argues that exploring these texts'

This chapter is a reworked version of a part of Chap. 2 of my PhD dissertation (see Ulstein 2021b).
When working on parts of this chapter, I received funding from the European Research Council (ERC) under the European Union's Horizon 2020 research and innovation program (grant agreement No 714166).

G. Ulstein (✉)
Centre for Literary and Intermedial Crossings, Vrije Universiteit Brussel, Brussels, Belgium
e-mail: Gry.Ulstein@VUB.be

B. H. Onishi, N. M. Bell (eds.), *The Call of the Eco-Weird in Fiction, Films, and Games*,
https://doi.org/10.1007/978-3-031-77126-2_10

"bad" or "irreverent" attitudes toward various aspects of the environmental crisis encourages "a less strictly instrumentalist approach" to ecology that "allows us to imagine additional, or different, capacities for environmental art—say, revealing the strict codes of environmentalism, expressing dissatisfaction or *disaffectation* with the environmentalist status quo, bearing witness to crisis, enacting catharsis, raising activist morale, building community, serving as cultural diagnoses, indexing and helping us understand our current eco-political moment, [...]" (2018, p. 7). According to Seymour, it is exactly when climate change-related issues become too real and too close for comfort that artists "reach for responses such as irony, absurdity, perversity, and camp" (p. 12). And because the strict codes of environmentalism often disfavors such responses, Seymour suggests it is vital to study artistic works that "embrace doubt, uncertainty, and ignorance as both rhetorical and aesthetic practices" (p. 13); that "emphasize the nonnormative and refuse to route things back to the familiar" (p. 33).

Seymour takes umbrage with the (eco)critical tendency to valorize certain affects, reactions, and themes as better suited or more valuable for ecological thought than others. Such implicit moralism permeates ecocritical discourse, Seymour shows, and pushes aside other, less palatable ways of responding to environmental crises. While documentaries like Netflix's *Our Planet* (2019) have a clear agenda to educate and to evoke in viewers a sense of care and grief for the (endangered) natural world (see Ulstein 2021a), artifacts that resist such blatant instrumentalism are worth paying attention to, writes Seymour, because they "give rare voice to the unclear, unexpected, or 'useless' feelings—the ambiguity, the ambivalence, the emotional stagnation—that crises like climate change trigger" (p. 72). Rather than allowing "the horrors of climate change, environmental racism, or factory farming to dampen their irony, their playfulness, their irreverence" (p. 234), these texts refuse to conform to normative expectations and, crucially, show that "engagement with serious issues need not entail serious affect or sensibility" (p. 232).

While "weird" is not among the registers Seymour mentions in her book, it fits the mold as a marginal aesthetic that contemporary artists gravitate toward. In this chapter, Seymour's project serves as a useful frame for my discussion of contemporary weird texts that challenge dominant (eco)aesthetic practices and (weird) genre conventions through discursive play, performativity, "bad" or "irreverent" weird affects, and other experimental narrative techniques—by which I mean "experimental" more generally or "experimental" for the weird specifically. In the famous essay "Supernatural Horror in Literature," H. P. Lovecraft suggests that the "whimsical and humorous" has no place in the weird tale, because "the author's knowing wink removes the true sense of the morbidly unnatural" (Lovecraft 2013, p. 446). True cosmic dread, for Lovecraft, must be "expressed with a seriousness and portentousness becoming its subject." Conversely, contemporary weird fictions increasingly engage the author's knowing wink via modes like metafictionality and self-reflexivity and experiment with affects not traditionally associated with the weird—like humor, irony, and love. In part, this might have to do with the weird's evolution into longer forms and different media than the short story and prose; (graphic) novels, films, and video games are often better suited for exploring and entertaining complex sets of affects and open up different avenues for experimentation.

The weird's increased formal experimentation might also, in part, be a reflection of the historical situatedness of the weird renaissance post-1990, at a time when postmodernist techniques such as metafictionality and self-reflexivity were well-established in literature more broadly.[1] But regardless of *why* weird texts started being more experimental, this tendency deserves more attention from scholars interested in the "eco-weird" because texts that engage in such narrative play to build ecological themes display a dual, interconnected resistance to traditional weird poetics on the one hand and conventional ecological thought on the other. On one level, as I will demonstrate below, these techniques target the reader's positionality and ethics vis-à-vis the text and, by extension, vis-à-vis the sociopolitical and ecological context. On another level, the experimentation also targets the weird itself, self-consciously questioning and stretching the way in which weird artists represent the weird's central trope: the cosmic fear of being human in uncertain times.

Before I continue, I should note that while ecocriticism is not quite "mainstream" within weird scholarship, it is not exactly new, either, to suggest that weird, weirding, or weirdness can provide a useful critical lens for discussing the global environmental crisis. Nor is it particularly groundbreaking within the environmental humanities to say that climate change and its many effects are "weird." The blossoming of a growing scholarly interest in the weird corresponds roughly with the birth of the discourse centered on the confrontational neologism "Anthropocene": the

[1] Mark Z. Danielewski's *House of Leaves* (2000), for example, is a weird text that frontloads such techniques.

geological epoch during which human influence on the earth's environment has become statistically significant in relation to other geological agents (Waters et al. 2016, p. 137). Alongside "global weirding," "weird realism," "Chthulucene," and "tentacular thinking," concepts like "dark ecology," "monsters and ghosts of the Anthropocene," "Anthropocene horror," "weird narratology," "weirded world," "weird ecology,"[2] and variations on these have grown in circulation between 2010 and 2024 and stretch across several philosophical corners of environmental thinking (in the examples above ranging from object-oriented ontology, critical theory, and new materialism to ecocriticism and narratology). Writers, readers, activists, artists, critics, scientists, and pedagogues from a broad array of disciplines (largely within the humanities) are turning to the weird to describe or critique the Anthropocene moment. Arising from this context, the present edited volume launches the eco-weird as a useful umbrella concept that might help synthesize and analyze the responses.

The growing interest in weirdness during our crisis-ridden contemporaneity has to do with the weird's setting in "unnerving edgelands" (Luckhurst 2017, p. 1056): porous storyworlds that subversively comment on how unstable the real world is and, by extension, how vulnerable human ontologies are. Within the environmental humanities and weird scholarship this commentary has often been interpreted as earnestly environmental or even explicitly activist. The deep horror mixed with wonder when faced with various forms of nonhuman entanglement in weird texts easily becomes an analogy—if a slippery, fuzzy one—of the ambivalent horror experienced when confronted with contemporary humanity's unfairly distributed entanglement in "hyperobjects" like biodiversity loss, global warming, or mass emergency migrations due to localized environmental disasters (Morton 2013, p. 61).[3] But the value of tapping into the weird as a resource for considering, understanding, and critiquing Anthropocene issues also has to do with the capacity of the weird for more

[2] In order of mention, see Canavan and Hageman, *Global Weirding* (2016); Harman, *Weird Realism* (2012); Haraway, *Staying with the Trouble* (2016); Morton, *Dark Ecology* (2016); Swanson et al., *Arts of Living on a Dying Planet* (2017); Clark, "Ecological Grief and Anthropocene Horror" (2020); Hegglund, "Unnatural Narratology and Weird Realism" (2020); Vermeulen, *Literature and the Anthropocene* (2020); Economides and Shackelford, "Weird Ecology" (2021).

[3] Morton coins "hyperobjects" in *The Ecological Thought* (2011), but more succinctly defines them in *Hyperobjects* as "things that are massively distributed in time and space relative to humans" (2013, p. 1).

than just earnest analogies to or profound insights into the status quo. There is also room for ambiguity, absurdity, and impropriety: there's room for playing with attitudes and reactions to crisis at the level of form as well as at the level of content. In *Bad Environmentalism* Seymour asks: "What does it mean when a work employs the 'wrong' tone or adopts the 'wrong' disposition for the subject matter in question [the environment]?" (2018, p. 21). Using Seymour's bad environmentalism as a springboard for my argument, in what follows I want to suggest that weird texts that play with narrative strategies, like affect, self-reflexivity, and metafictionality, deserve attention not just because of these texts' well-established fearsome fascination with (hyper)objects that evoke a sense of "wrongness" (Fisher 2016, p. 16). These texts also deserve attention because they embrace traditionally "wrong" weird narrative techniques, coming at ecology from new angles that offer other, perhaps less correct, ways of responding to the Anthropocene and its many crises.

Like Seymour, I have selected a few vivid examples that work particularly well to argue this point. In the first part of the chapter I will take a moment to elaborate on the role of humor and comedy in the weird and in ecocriticism before I turn to Jeff VanderMeer's novel *Borne* (2017) to discuss experimentation with affects such as humor and love—focusing on humor because the arguably thin line between horror and humor gives *Borne*'s ecological themes a critical edge. In the second part I will explore how these themes are taken to a more complex level in *Cyclonopedia: Complicity with Anonymous Materials* (2008), in which Reza Negarestani uses explicitly experimental techniques like self-reflexivity, intertextuality, and humor as intertwined, subversive strategies to build a case against everything the weird hates: anthropocentrism, capitalism, and passive consumer-readers. To start, however, I will briefly discuss a short text in which humor is pointedly weirded in order to offer ecological critique.

10.2 Weird Humor as (Ecological) Resistance

Shaun Tan's short story "Bears with Lawyers" (2020b) starts with a speculative punchline for the Anthropocene: "Bears with lawyers. It was as simple and terrible as that" (125). The barely four-page story scaffolds a world in which bears are suing *Homo sapiens* for their crimes against *Ursidae*, but all that is needed to activate the environmental imagination of the reader is really those two first sentences—even without Tan's absurd illustration accompanying the story (Tan 2020a). The fantastical elements reinforce the satirical humor and vice versa, like when the bears are done listing the many crimes for which they are suing ("Human Law is not the only legal system on the planet, it turned out"), and conclude with a scathing remark: "'For the hungriest of all animals,' said the bears in their typically abstruse legal verse, 'the only thing left to eat is the truth.' As if to prove their point, none of our lawyers could view the supporting library video of evidence without losing their lunch" (2020b, pp. 125–126).

The human we-narrator represents humanity as one hypocritically indignant voice, with the exception of the bears' human lawyers who say they "had no choice" but to take the case because they are "sworn to uphold justice" (Tan 2020b, p. 128). The satire already distances this human collective from the reader somewhat, and it becomes obvious where our sympathy is directed as the bears point out that the human systems of economy and justice mean nothing compared to "the Ledgers of the Earth, written in clouds and glaciers and sediments, tallied in the colours of the sun and the moon as light passed through the millennial sap of every living thing" (p. 127). This sentimental insight into nonhuman systems of value almost overpowers the humorous tone of the story as humans are confronted with the fact that, compared to the Ledgers of the Earth, all they own is capital, which is nothing but "a collective figment of the human imagination" (p. 127). But the dark humor returns full force as the humans, true to form, simply shoot the bears to deal with the problem so that "finally, finally, finally, things went back to normal" (p. 128).

But as Gerry Canavan and Andrew Hageman have pointed out, "we are now living in postnormal times," in which "we can no longer depend on the climatological patterns that up till now have more or less reliably structured our behaviors, including our architectural, agricultural, and resource-extractive practices, as well as the life patterns of the plants and animals with which our coexistent surviving and thriving depends" (2016, p. 8). Canavan and Hageman use the above as part of their rationale for adopting the term "global weirding" as an alternative way of addressing the warming world and its (arguably) weird effects. Likewise, Tan's story takes a more sinister turn to the weird as it ends with humans being "served" by a client much worse than bears: "Blood is leeching from every face in the boardroom, and we can already hear the sound of document trucks reversing into streets. A fetid cloud descends over the city, the sickening stench of endless torment and persecution. 'The cattle are here,' a terrified receptionist quavers over the intercom, 'with lawyers'" (Tan 2020b, p. 128).

While "Bears with Lawyers" might resist categorization as a weird story because of its relative bluntness, in a way analogous to the concept "global weirding" it is a tongue-in-cheek example of how humor and absurdity can be used alongside horror and weirdness to offer ecological critique. In Tan's story, the last paragraph is probably where weirdness seeps in most clearly—particularly the "fetid cloud" and "sickening stench of endless torment and persecution" in the wake of the cattle—and yet, the humorous double meaning of being "served" (by) cows contrasts this weird charge. But the story's main source of horror can be found in the implicit point of its dark joke: no matter how terrible the fury of the cows, the true horror, for all nonhuman species, is humans. In the age of the Anthropocene, humans must come to terms with and take responsibility for their geological agency or accept the consequences of a world turned against humans, Tan suggests. It is a court case they cannot possibly win. This kind of dark humor is not unknown to weird fiction. The premise of *Roadside Picnic* (1972) by Arkady and Boris Strugatsky is that aliens visit Earth but depart again before anyone notices; having left behind a bunch of weird junk with sinister powers, ostensibly after deciding Earth was not interesting enough to warrant more than a quick stop. In China Miéville's short story "Covehithe" (2015), the punchline is oil rigs that come alive and crawl up on beaches around the world to lay eggs like giant turtles, implicitly asking what happens to petroculture in the wake of this transformation. Jeff VanderMeer also plays with a similar darkly funny poetic justice in *Dead Astronauts* (2019), when a character called the blue fox and his army attack and murder humans with the humans' own chainsaws to teach them a lesson about cutting down trees.

But as mentioned above, humor in weird fiction has not received much critical attention. This is perhaps not very surprising, considering the rather unfun topics of cosmic indifference and imminent apocalypse that tend to dominate the weird. Such frightening themes also, in part, explain the weird's relative popularity with an Anthropocene readership and its suitedness for ecocritical analysis. However, in contemporary weird narratives there are scenes or techniques that evoke both dark and more lighthearted or whimsical humor at different diegetic levels. This humor can function as a mere contrast to the horror, it can enter into the discourse via implicit or explicit satire as in the examples mentioned above, or it might work to subvert the entire weird aesthetic. Of course, traditional weird tales might evoke comedy in a different sense.

Mark McGurl proposes "posthuman comedy" as a way of addressing issues of deep time and large scales by looking to "those literary forms willing to risk artistic ludicrousness in their representation of the inhumanly large and long" (2012, p. 539). For McGurl, significantly, this means looking to genre fiction like science fiction and horror, and more specifically to Lovecraft's weird, a literary tradition which historically has been "utterly beyond the pale of artistic seriousness" (p. 543). Refreshingly, and in a way that runs counter to most engagements with Lovecraft, McGurl takes the weird, as a representative of genre fiction, "seriously" by situating it outside the institutionalized canon and boasting this position as a strength:

> Not only does genre fiction seem to violate the law of writing what you know from personal experience; not only does it bear its "formulaic" flatness on its grubby sleeve, catering to tastes unformed by the university, but its darkly dorky aesthetic unseriousness is an affront to the humanities—hell, an affront to humanity. Look at those characters, little more than the toys of allegory! If only genre fiction exhibited the chastity of quantitative representation one finds in a scientific paper; but, no, it insists on the comic personification of the absolutely other. (McGurl 2012, p. 550)

Through this insistence on the "comic personification of the absolutely other," McGurl suggests that weird fiction is in a unique position to laugh at human insignificance—not, he writes (referencing Henri Bergson's essay *Laughter: An Essay on the Meaning of the Comic*), simply by being funny but via "the broader and frequently unfunny 'comedy' of the human condition such as we find it in Dante's *Divina commedia* or Honoré de Balzac's *Le Comédie humaine*, where it is associated with various forms of ontological *lowliness*" (2012, p. 549, emphasis in original). This connection, via genre history and ideas about "high and low" literature, between genre fiction and posthumanism, technology, and deep time places the weird in a strong position for Anthropocene critique by scrutinizing its capacities for ridiculing the human as part and parcel of its cosmic horror aesthetic.

McGurl's view of the weird also recalls Seymour's bad environmentalism, where she discusses "self-parody and self-criticism" as qualities that contrast "the sanctimony and self-righteousness found in much environmental art, activism and discourse" (Seymour 2018, p. 145). Perhaps I am doing Seymour's project a disservice by suggesting that Lovecraft's infamous purple, and often racist, prose can be used to do the same important work of "emphasiz[ing] the nonnormative and refus[ing] to route things back to the familiar" (2018, p. 33) as Seymour's primary texts. Seymour is in fact suspicious of humor and of comedy as an ecocritical mode in general, because "traditional generic definitions of comedy refer to works that offer reconciliation and resolution" and "restore normalcy" (2018, p. 33). But as I will demonstrate further down, humor in the weird is not simply a loophole to normalcy. Rather, its occurrence plays with and complicates the weird aesthetic and, by extension, the ecological themes explored. In McGurl's discussion of Lovecraft, of course, the humorous effect of the weird comes from a higher-level reflection on the foolish human: weird fiction, for all its horror, also laughs "darkly dorky" at the human. Indeed, it is clever of McGurl to propose taking weird fiction seriously as literature by explicitly *not* taking its cosmic horror seriously, but reading it instead as a commentary on human failings. Or as McGurl humorously puts it, borrowing a famous phrase from "The Call of Cthulhu": "Excluded from spaces of literary institutional respectability, Lovecraft sought to visit the horrible truths of deep time upon the island of ignorance formed by the conventions of literary realism" (McGurl 2012, p. 545).

As McGurl suggests, the weird therefore stands to contribute to ecocritical thought not just through its cosmic horror but also through its overarching posthuman comedy. Most evocatively, McGurl meshes Lovecraft and evolutionary biologist Lynn Margulis to make the point that it is worth paying attention to narratives that stage the human as comical or insignificant in relation to imperceptible scales (massive as well as microscopic):

> For Margulis, human beings are simply not the "dominant species" they take themselves to be but rather the clown, the "planetary fool" dressed up to deny his deepest identity as "glorified sludge." And it's true that for every gelatinous green immensity we could hope to see, there are untold legions of insects and mollusks, seething billions of microbes, trillions of atoms not less disturbing, from a certain perspective, to the dignity of humankind than old Cthulhu. (McGurl 2012, p. 552)

McGurl's careful discussion of historical perspectives on narrative and the human in deep time culminates in a comparison between Lovecraft's loquacious "gelatinous green immensities" and American writer Raymond

Carver's minimalist "slices of life": according to McGurl, these very different writers intimate the same posthuman comedy (2012, p. 553). The comic human McGurl identifies in Carver, Margulis and Lovecraft is not humorous so much as it is tragi-comical, as the locus of the posthuman comedy is in the cosmic-situational ridicule of being human rather than in any kind of humor affect. Nevertheless, McGurl's engagement with Lovecraft places the weird in the position of making fun of humanity next to its position of evoking pure, serious, cosmic horror—while simultaneously taking genre fiction seriously where other critics might not. To expand McGurl's point, it might be argued that the traditional weird invites the reader to be entertained by a form of unintentional humor precisely because it takes the ridicule of humanity so seriously. Thus comedy enters into the traditional weird at the discourse and meta-critical levels while remaining absent at the level of the storyworld.

The lack of interest in comedy/humor and the weird might thus have more to do with (the lack of) critical interpretations and serious engagements with the weird than with weird fiction itself. VanderMeer has included a brief note on his personal blog where he claims that some weird stories (his examples are the authors Thomas Ligotti and Angela Carter) do have a kind of "dark humor that stands out in relief because one element of the story is slightly less perverse than the rest" (VanderMeer 2010, "Weirdly Funny"). It might be pushing it to argue that one element being "slightly less perverse" constitutes humor—but as Noël Carroll has argued: the presence of category errors or inflated contradictions combined with other textual cues can quite easily tip the scale from horror to humor (Carroll 1999, p. 152).[4] In "Bears with Lawyers" this reliance on category errors is abused to its fullest to support Tan's satire. But in spite of a substantial amount of work on dark humor in other genres and media, only minor remarks and notes similar to VanderMeer's make up the critical engagement with humor and the weird. And interestingly, this relative critical disinterest in humor also marks the field of ecocriticism.

Courtney Traub concludes an essay on humor and climate fiction by stating that "ecocritics have only recently begun to consider the place of humor and satire in environmentally themed fictional writing" (2018, p. 104). This is a rather striking conclusion, as Traub also points out,

[4] Carroll suggests that horror, just like humor, is "intimately and essentially bound up with the violation, problematization, and transgression of our categories, norms, and concepts" (1999, p. 152)—they are merely followed by a differently anticipated emotional reaction.

considering the legacy of dystopian, eco-oriented satire represented by influential works like Aldous Huxley's *Brave New World* (1932) or Kurt Vonnegut's *Slaughterhouse-Five* (1969), and in light of popular works of climate change fiction that use humor actively, such as Margaret Atwood's *MaddAddam* trilogy (2003–2013) and Ian McEwan's *Solar* (2010)—the latter two are discussed closely in Traub's essay. In the essay, Traub—like McGurl—argues that genre fiction deserves ecocritical attention, and she echoes Seymour, who calls out ecocritics and environmentalists for being overly earnest and self-righteous in their critique. For this reason, Seymour is more interested in "works that go beyond or complicate the modes of humor and the comic" (like irony, queer, absurdity, and perversity) because humor and comedy can also "reaffirm normative values" (Seymour 2018, p. 33). Humor and comedy, for Seymour's project, easily acquire unhelpful overtones of sanctimony and moralism, as demonstrated by, for instance, the much-discussed Netflix movie *Don't Look Up* (2021). When it comes to the weird, however, the texts that experiment with humor to create a deliberate tension with horror display a similarly dissident stance toward weird genre conventions as Seymour's "often-denigrated affects and sensibilities" (Seymour 2018, p. 6). Likewise, drawing on the incongruity theory of humor,[5] Traub suggests that "[w]hen presented with topics as dire as anthropogenic climate change, species extinction, or biochemical warfare, the laughter or amusement unexpectedly elicited by their humorous treatment disrupts an expectation that such subjects be presented in a wholly serious register" (Traub 2018, p. 91).

Traub's claim about the ecocritical disinterest in humor is not necessarily in conflict with Seymour's dismissal of humor as an affect, as Seymour's corpus does not include speculative fiction. At its worst, as Seymour reminds us, humor can "enact the same problematic dynamics and perceptions that have long plagued mainstream environmentalist movements" (2018, p. 57). At its best, however, in weird fiction and other types of genre fiction, humor can function as a counterweight to horror: complicating, personalizing, or even subverting it, whereby it presents an

[5] See John Morreall's "Philosophy of Humor" in *The Stanford Encyclopedia of Philosophy* (2020), which traces the incongruity theory back to eighteenth-century thinkers like Kant, Schopenhauer, and Kierkegaard and their critiques of the "Superiority Theory" of humor: "While the Superiority Theory says that the cause of laughter is feelings of superiority, and the Relief Theory says that it is the release of nervous energy, the Incongruity Theory says that it is the perception of something incongruous—something that violates our mental patterns and expectations."

"alternative means by which environmental risks might be represented and ethically engaged" (Traub 2018, p. 104). Before I turn to Negarestani's *Cyclonopedia*, where humor is only one of several such alternative means, I will discuss VanderMeer's novel *Borne* as an example of a weird eco-fiction using non-weird affects like humor actively to contrast and heighten, but also disrupt, the horror of the Anthropocene.

Borne is part of a sort of trilogy and set in the same storyworld as the novella *The Strange Bird* (2018), which is experimental in its use of non-human focalization, and the novel *Dead Astronauts* (2019), which is the most experimental of the three in terms of form and metafictionality.[6] *Borne* is not particularly experimental in terms of form, with a straightforwardly eco-oriented thematic and a largely linear plot structure narrated from the scavenger Rachel's first-person perspective. The novel is set in a postapocalyptic storyworld where biotechnologically enhanced creatures roam the wasteland left behind after decades of war and corporate corruption. For the most part, *Borne* builds suspense via exciting scenes that suggest secrets and conspiracies, painting Rachel as a resourceful, clever, and intrinsically good person—and the biotech, uncanny wunderkind Borne as a catalyst for positive growth on the one hand and catastrophic destruction on the other. Many of the scenes in which Rachel and Borne, the two main characters, are in danger emphasize and strengthen the bond that has formed between them across species boundaries, which endears them both to the reader. By the end, after Borne has sacrificed himself to kill the malicious monster Mord, the reader is left with closure and resolution, despite the dystopian setting.

However, the novel's use of humor and other un-weird affects goes a lot further than making elements of the story "slightly less perverse," as VanderMeer suggests in his comment about humor in weird stories. Carroll notes that "where the fearsomeness of the monster is convincingly in place, horror will not drift over into incongruity humor. But where the fearsomeness of the monster is compromised or deflected by either neutralizing it or at least drawing attention away from it, the monster can become an appropriate object for incongruity humor (Carroll 1999, p. 175). This insistence on the monster as a defining characteristic of

[6] See my paper "'Just a surface': anamorphic perspective and nonhuman narration in Jeff VanderMeer's *The Strange Bird*" (Ulstein 2021c) and the paper I co-wrote with Marco Caracciolo, "The weird and the meta in Jeff VanderMeer's *Dead Astronauts*" (Caracciolo and Ulstein 2022).

horror has obvious shortcomings if we consider twenty-first-century horror, and particularly the weird: there isn't always a monster to fear. Even when there is a monster, the monstrous is often complicated by degrees of monstrosity or by modes and affects beyond both horror and humor, such as hope or grief, or, as VanderMeer explores elaborately next to humor in *Borne*: love. Rachel raises Borne as if it were her child and comes to care for him with a comparable passion. *Borne* moreover plays with the incongruity slip between horror and humor by personifying it in the novel's two "main monsters": Borne and Mord.

Rachel finds Borne entangled in the fur of Mord: the great, flying bear who torments the city ruins where Rachel and her partner Wick live as scavengers. At first glance, there isn't much in the setting that invites humor. As Rachel herself reflects, the bleakness of the storyworld makes monsters of them all: "In the city, the line between nightmare and reality was fluid, just as the context of the words killer and death had shifted over time. Perhaps Mord was responsible. Perhaps we all were" (VanderMeer 2017, p. 28). But the scenes where Borne and Rachel interact quickly become recognizable as the lighthearted, often humorous relationship between parent and child or human and pet. As I have argued elsewhere,[7] the scenes where Rachel teaches Borne new words are particularly humorous and whimsical, and the motif of language is important for sustaining the use of humor throughout the novel:

> "What rhymes with *crappy*?" Borne would ask.
> "*Happy*?"
> "No, *shitty*."
> "No, that word doesn't rhyme with *crappy*."
> "But it rhymes with *city*, and that rhymes with *happy*."
> "None of that is true."
> "*True* rhymes with *fact*."
> "In a way, I guess."
> "*Fact* rhymes with *city* and *happy*."
> "No, in this case *city* and *happy* put together rhyme with *opinion*."
> "You don't share my opinion?" (VanderMeer 2017, p. 90)

This attempt by Borne at manipulating Rachel into taking him into the city is an innocent, funny counterpart to his more alarming ability to physically manipulate matter, as Borne's natural shapeshifting into other

[7] See Ulstein (2019).

creatures can only be perfected if he kills them first. Rachel willfully ignores this for as long as she can, but in the end, the love Rachel feels for Borne is challenged by his sinister urge to absorb and by his questionable (non-human) ethics concerning other creatures. The scene where Rachel catches Borne in shapeshifted form looking like her leaves her wondering if Borne has ever put on her or her partner's shape when they were in bed together (VanderMeer 2017, p. 182), an incestuous implication which grants Borne another level of monstrosity. In other words, while its evocation of love might run counter to affects explored in traditional weird fiction, the novel does not surrender to it as a form of easy escape, either. The presence of love in *Borne* is also complicated by the fact that Rachel's attempted parenthood fails, as Borne's alterity is too much for her in the end and she casts him out.

And yet, there is something cuddly and cute about Borne—something reminiscent of the iconic alien in Steven Spielberg's *E.T. the Extra-Terrestrial* (1982) or the rubbish-cleaning robot in Andrew Stanton's *Wall-E* (2008)—which might raise suspicion in readers familiar with the weird. "When Borne saw me staring at him," Rachel notes a few weeks into her adoption of him,

> he would make a sound like the startled clearing of a throat, and his flesh would absorb all of the eyes except two, which would migrate higher on his body and away from each other. Sometimes they would slip back down to his hips, but once in position on his torso they became larger, took on a sea-blue color, and grew long, dark, lashes; they moved independent of each other. He must have thought he looked more normal that way. (VanderMeer 2017, p. 44)

The description of Borne's many vertiginous bodily attributes is simultaneously estranging and endearing, but the attentive reader suspects that the sense of endearment will soon evaporate. Even before the true horror of Borne is revealed (his ability to absorb, copy, and kill everything he touches), readers are primed to expect it because of the mere incongruity of cuteness and humor in a story that otherwise cues horror and gloom. This tension is played with throughout *Borne*, bringing relief from the dreary setting while also building suspense and moreover demonstrating the weird's growing "sensibility of welcoming the alien and the monstrous as sites of affirmation and becoming" (Noys and Murphy 2016, p. 125). In *Borne*, as in most of the eco-weird, this welcoming is messy and complex and relates to the central question the novel poses: how do you exist as a(n ethical, good) person in a world that has changed so much for the worse?

While Rachel is busy raising Borne to be a “person” (a concept Borne keeps questioning and challenging), her partner Wick is hiding the fact that he, like Borne, is biotech: not human, but merely mimicking humanity. Toward the end of the novel, when Wick’s identity has been revealed, Rachel concludes: “Wick never believed he was a person, was continually being undone by that. Borne was always trying to be a person because I wanted him to be one, because he thought that was right. We all just want to be people, and none of us know what that really means” (VanderMeer 2017, p. 320). *Borne*’s exploration of personhood extends beyond the characters in the novel and beyond humanity in general. In the wake of human-caused destruction at the headquarters of the evil Company, all kinds of animals have “gotten in through the cracks,” Rachel notes, and VanderMeer here facetiously references one of Lovecraft’s short stories: “The rats in the walls, who were in the process of rewiring everything. They were the future” (VanderMeer 2017, p. 297). This posthuman outlook into a future no longer dominated by humans is a recurring theme in VanderMeer’s work (particularly in the Southern Reach trilogy). In McGurl’s sense of the posthuman comedy, there is something comic about this prospect, with humanity becoming the butt of a joke they never understood before it was too late, before the world changed so much that they would start paying the price for their lack of understanding. In Tan’s short story, this joke is played out dryly, mercilessly, and extradiegetically: the humorous affect arises mainly from the we-narrator’s indignant horror, which the reader knows, and the author knows the reader knows, is deserved. In *Borne*, the humor is evoked intradiegetically as well, as in the familial conversations between Rachel and Borne, but also in several of the scenes portraying Mord.

The mere fact of Mord being a bear—a flying bear, no less—with human characteristics and moods, carries a lot of latent humor, much like the bears in “Bears with Lawyers.” Like Borne, Mord has more elaborate moments of comedy too, but the comical aspects of Mord are indicators that the horror he represents is no longer the same as the cosmic horror of the traditional weird. For example, in a scene where Rachel and Borne are resting on a rooftop, looking out over the ruined city and the night sky as Rachel explains to Borne what stars are and how the earth revolves around one, suddenly “something began to blot out the stars, to turn that glittering, shining brilliance into a great and final darkness" (VanderMeer 2017, p. 110). At first, Rachel is "speechless" because she thinks "the world was ending," but when she realizes it is Mord, she "couldn't help a stifled chuckle. Oh, this was rich! Because it was the end of the world" (p. 111). When Rachel locates the source of the "final darkness" and realizes it is Mord hovering over them, it loses its power to awestrike her. A few pages later, as she is chatting with Borne, "Mord had gone in our estimation from 'spectacular' to 'buffoonish,' his star-blotting the work of a clumsy, maniacal, *floating* bear. 'Whoever heard of a floating bear!' I told Borne. 'That'd be like finding a plant that was actually a talking octopus'" (p. 113). The last sentence is doubly humorous because it caricatures Borne as he appeared to Rachel when she first found him.

In other words, even as we, as readers, are told that Mord is terrible, impossibly large, and has incomprehensible motivations, it does not quite cover the fact that he is just "a clumsy, maniacal, *floating* bear." Mord's bluntness seems a deliberate caricature of Lovecraft's incomprehensibly massive monsters—all the more so because everything is focalized through the perspective of Rachel. Lovecraft's perpetually male, rational, elitist scientist who is driven mad by the mere mention of Cthulhu is dismissed and replaced in *Borne* by a female climate refugee and scavenger who tries (not always successfully) to respond to Mord as nothing more or less than he is and whose engagement with Borne (although it scares her and puts her in danger) is deeply personal. While *Borne* does not seem to argue that the "new" chimeric, personhood-keen monsters like Borne and Wick are necessarily better, the novel does suggest that monsters like Mord and Cthulhu have become conventionalized and comical, their alleged horror too blunt and disinterested to successfully represent the many facets of terror faced in the Anthropocene. By extension, Anthropocene issues are also caricatured. When Borne asks what happened to the world, Rachel lists a series of platitudes in her head: "We didn't try hard enough. We were preyed upon. We had no discipline. We didn't try the right things at the right time. We cared but we didn't do. Too many people, too little space" (VanderMeer 2017, p. 166), but saying out loud after a "scathing" laugh: "That's just the way it is, Borne. Survival isn't pretty" (p. 167). And this blunt, caricatured environmental critique—similar to the bluntness of Mord's horror—does not read as a termination of ecological thought or a bleak dead end but rather as a challenge to find alternative ways of looking at it.

Borne thus responds to one of Seymour's central questions in *Bad Environmentalism*: "How are we *supposed* to feel in our relations with environment and living creatures, [and] what happens when we do not feel that way?" (Seymour 2018, p. 21). There is space in the eco-weird, as there is in environmentalism, for unexpected, rebellious affects, moods, and perspectives: for wrongness and play. Addressing this dissident spaciousness expands the eco-weird's critical potential and acknowledges one of the central ways in which the weird has changed as a genre and mode in the twenty-first century. *Borne* is a great example in this regard, as it embodies the evolution from the short to the long form, the thematic shift from the cosmic to the political and the personal, and the inclusion of contrasting affects like humor and love. It is important to note that the slip between horror and humor and love does not exhaust Borne's mix of registers and affects, which includes mourning, nostalgia, despair, and hope. But by using humor and love as explicit counter-affects to the cosmic horror found in traditional weird narratives, VanderMeer transgresses generic boundaries and implicitly corroborates the irritating sanctimoniousness that Seymour, McGurl, and Traub identify in ecocritical discourse. While humor as resistance remains important for the second part of this chapter, this part will center on a text that showcases the eco-weird's capacity for complex formal and thematical experimentation beyond affect.

10.3 Planetary Plot Holes as Weird Play

As I have argued so far, un-weird affects like humor and love can disrupt conventional weird narrative structures and, through this disruption, subvert anthropocentrism or challenge traditional ways of thinking about the Anthropocene. The texts I have mentioned or discussed, even though they experiment with weird genre conventions, are not outspokenly experimental beyond such affective and thematic disruption. Moving on, Negarestani's *Cyclonopedia: Complicity with Anonymous Materials* troubles ideas about literature to the extent that discussing it in terms of genre or affect is in danger of diminishing Negarestani's project. In the words of one critic: "the book's prose resembles what would happen if Deleuze and Guattari collaborated with H. P. Lovecraft" (Fest 2016, p. 571).

Negarestani's book is at once thought experiment, weird tale, political satire, and reading strategy for twenty-first-century oil-related crises.[8]

Cyclonopedia takes weird tropes and techniques and twists them into stories that are only stories because readers bring that expectation to the text. By playing with metafictional strategies and exploiting the tension between horror and humor, *Cyclonopedia* demands from readers that they consider its gaps and absences (or "plot holes") as much as its contents. In a sense, Negarestani offers a satirical horror version of Edward W. Said's method of contrapuntal reading from *Culture and Imperialism*, which adopts "a simultaneous awareness both of the metropolitan history that is narrated and of those other histories against which (and together with which) the dominating discourse acts," so that "extending our reading to include what was once forcibly excluded" can "take into account both processes, that of imperialism and that of resistance to it" (Said 1993, p. 51; pp. 66–67). Where Said's dominant narrative process is imperialism, however, Negarestani's are capitalism and petroculture—three processes, of course, that are all inextricably intertwined. Purposefully unclear about good alternatives to these processes, Negarestani does propose various ontological and aesthetic strategies for resisting them, such as Deleuze and Guattari's politics of becoming (Negarestani 2008, p. 196), "radical openness" (p. 201), and the acts of engaging in "Hidden Writing" and reading through "plot holes" (p. 60–64), which I will discuss in more detail below.

Readers of *Cyclonopedia* are warned about the challenging reading experience to come in Kristen Alvanson's journal notes prefacing the "thick piece of writing titled *Cyclonopedia* with the name Reza Negarestani handwritten on it" (Alvanson 2008, p. xii), which she claims to have found under her hotel bed in Istanbul: "Sections each paperclipped with various articles for each chapter, some pages nearly hidden under small yellow post-it notes or with margins saturated by handwritten notes and drawings in black pen and pink highlighter, making it nearly impossible for me to read" (p. xii). This *mise en abyme* evokes the well-traveled weird trope of the found manuscript but immediately targets its own (in)authenticity when Alvanson next quotes a passage she "finds" scribbled in the margins of the book:

[8] Or as *Cyclonopedia* self-identifies on the back cover: "At once a horror fiction, a work of speculative theology, an atlas of demonology, a political samizdat and a philosophical grimoire" (Negarestani 2008).

> Xenopoetics has something to do with composing out of distorted materials. One page is missing, one or two lines are pseudonymously or anonymously quoted, one scene leaks from the future to the past, an object evades chronological sequences, a number turns into a cipher, everything looms as an accentuated clue around which all subjects aimlessly orbit, leading into an eclipsed riddle whose duty is not to enlighten but to make blind (aporos to the light). (p. xvii)

The passage above, conversely, is nowhere to be found in the printed manuscript *Cyclonopedia*, which Alvanson claims to have taken "back to the US with the intention to publish it" (p. xx). The other notes in the margins *have* been transcribed and indicated with a double dagger symbol at the bottom of the page and might or might not have been authored by Negarestani. They contain research or reflections on the process of writing *Cyclonopedia*, allegedly in collaboration with a number of anonymized people (S, Ƨ, Z, K, A, M, and B), each of whom "seems to disappear without a trace" (Alvanson 2008, p. xvi).

Cyclonopedia, it quickly becomes clear, *is* xenopoetics—or "poetics of the Other/Stranger": it involves all the fuzzy qualities mentioned in Alvanson's spectral margin definition above. At this point in the text, readers will be tempted to do what Alvanson does on the following page: look up Reza Negarestani online. Any search will reveal that Negarestani is indeed a flesh-and-blood Iranian philosopher born in 1977 and with several published books, although Alvanson is dubious as to the legitimacy of this information and reviewers tend to reinforce this conspiracy theory by reporting that "debates have raged on as to whether he actually exists" (Blas 2012a, p. 10). Further reviews, blog posts, and academic papers suggest a whole collective of artists and academics who are in on the joke and, in a way similar to how readers of Lovecraft have perpetuated the myth of the grimoire called the *Necronomicon*, readers of Negarestani are keeping alive the "myth" of Reza Negarestani. At Alvanson's invitation—Alvanson herself a real-life artist and friend of Negarestani working in the Middle East—readers join a metafictional, mock-conspiracy game when engaging with *Cyclonopedia*, constantly confronted not only with the materiality of the text but with the possible, and humorous, *immateriality* of the author. This game, moreover, together with the weird slips between horror and humor both in the text and outside the text, becomes the catalyst for Negarestani's Anthropocene critique.

As mentioned, it makes little sense to characterize *Cyclonopedia* as either fiction or non-fiction, as blurring that boundary is the book's explicit ambition. The definition of *Complicity with Anonymous Materials* is listed as "inauthenticity" in the gappy, semi-helpful glossary (Negarestani 2008, p. 241), which signals that the project of the book is not so much to establish a clear-cut line between reality and storyworld, the actual and the virtual, as to suggest that the real is already subverted and dominated by the fictional and the virtual in many forms. Kate Marshall has noted that "*Cyclonopedia* has all of the trappings of a postmodern novel: a self-subverting metafiction that destroys its own grounds even as it enacts them" (2012, p. 154). According to Marie-Laure Ryan, such texts front-load play over immersion and keep readers "in a state of permanent jet-lag" because of the changing perspectives and the way they create "multiple realities that relativize each other" (Ryan 2001, p. 199). *Cyclonopedia*, however, extends its experimentation beyond self-reflexivity and instead suggests a text trapped in a kind of "videogame narration." For as Negarestani notes in the glossary: "Videogames enforce the logic of nightmares through their ludicosm" (Negarestani 2008, p. 244). The "ludicosm" of *Cyclonopedia* is a complex set of self-reflexive and intertextual theory essays, saturated with storytelling metaphors and obsessed with their own claim to materiality. Marshall astutely points out that *Cyclonopedia* goes beyond postmodernist texts because it invites "interaction instead of interpretation" by referring to websites, collectives, and real persons that have all contributed to the book and will continue to take part in the discussion (Marshall 2012, p. 154). If postmodernist texts keep readers in a permanent state of jetlag, *Cyclonopedia* induces a form of playful paranoia.

On a thematic level, *Cyclonopedia* is essentially an attempt to think through oil as an inflated object in the Earth's geological strata, which has been permitted to infiltrate human and nonhuman life to the extent that world processes cannot be comprehended without oil. The chapters are largely organized as scholarly discussions related to the findings and notes of the invented archeologist ("paleopetrologist") Dr. Hamid Parsani, who digs up the Cross of Akht: a "sentient relic with the ability to numerically grasp all the undercurrents and inconsistent events of the Earth as modes of narration" (Negarestani 2008, p. 13). According to Parsani, the Cross of Akht can *narrate* "the plot holes of every planetary scenario" by using oil as its lubricant (13). Understanding the Earth as a whole, for Negarestani, means to realize that it is operated or narrated via interconnected "plot holes" that constitute what Negarestani calls "Hidden

Writing." Hidden Writing is a reading strategy for the "perforated structures, degenerate formations and plot holes"; or those "perforations and anomalies in a text caused by the existence and activities of something other than the governing structure or the assumed plot" (Negarestani 2008, p. 60).

This idea of Hidden Writing is comparable to Said's contrapuntal reading, as mentioned above, but it also obviously harks to (and explicitly refers to) Deleuze and Guattari's minor literature and "becoming-minoritarian" in *A Thousand Plateaus* (1987). Becoming-minoritarian is an ethical position or act that, roughly summarized, actively subverts and challenges power structures and grand narratives (see Deleuze and Guattari 1987, p. 291). In Hidden Writing the "central or main plot is reinvented solely in order that it may stealthily host, transport and nurture other plots" (Negarestani 2008, p. 61), which reads at once as the "point" of *Cyclonopedia* and as a critique of conventional, plot-based narratives that tend to gloss over or disadvantage minor narratives. Engaging the mode of Hidden Writing thus opens the Earth's plot holes to scrutiny and critique, and the most glaring conspiracy running through this perforated Earth is oil—or in Negarestani's words: "hydrocarbon corpse juice," which "has been the source of ideology for petro-masonic orders and their policies—from OPEC to the agencies of War on Terror to promo-leftists ... extracted through topological instrumentalization of the socio-political body of the Earth" (Negarestani 2008, p. 27; ellipsis added). Oil, in *Cyclonopedia*, is the great antagonist that "turns Time toward apocalyptic blasphemies," hence the many Lovecraft references, because "[i]s there anything more Lovecraftian than the building of a new pipeline, winding its blobbing flutes?" (Negarestani 2008, pp. 58; 72).

Where Miéville's "Covehithe" critiques petroculture through cute oil rig babies that ostensibly eat all the oil before humans can get to it, oil in *Cyclonopedia* is the zombified blood of the Earth's nonhuman depths, dug up and forced to become the capitalist "corpse juice" of the human-dominated surface. Oil is given the status of "Tellurian Lube" in a humorous, whimsical way—even if the equation between "corpse juice" and "lube" is unpleasantly necrophiliac. But at the same time it is also deadly serious, because oil "lubricates" all kinds of areas of twenty-first-century human life, from monotheistic religion to consumerism, from the War on Terror to global warming. "To grasp oil as a lube," therefore, "is to grasp earth as a body of different narrations being moved forward by oil. In a nutshell, oil is a lube for the divergent lines of terrestrial narration"

(Negarestani 2008, p. 19). It is sometimes unclear whether, and how, "narration" is different from "structure" or "reality" in *Cyclonopedia*, except perhaps that "narration" emphasizes the constructedness *of* reality. In any case, one of the takeaways from the coveted Cross of Akht and its narration of Earth through oil is that readers—and here Negarestani echoes Seymour—too easily crave and accept wholesome structures, meaningful objects, or hermeneutical instruments that help them find (the) greater meaning in a text. Or, as Negarestani notes, such readings see "plot holes [as] tricks, they are ill-timed and ill-spaced coordinates within the text—leak holes which must be plugged" (p. 62). Negarestani conceptualizes the Earth as one big narrative machine driven by oil and accessed via myth, archeology, horror, and Deleuzoguattarian philosophy. Only in the plot holes of these intertwined, oily narratives can the conspirative connections emerge into the open—be "exhumed." The implicit, unanswered question, of course, is what will lubricate human life on Earth when the oil is gone.

The registers of *Cyclonopedia* thus shift between satire, horror, humor, and nihilism, all the while framing a radical form of new materialism,[9] with both inflated and obtuse critiques of institutionalized religion, (western) hermeneutics, and capitalism. Moreover, with *Cyclonopedia* Negarestani performs a defamiliarizing unmooring of the reader by exploring deep-temporal, fluid—*oily*—ways of viewing the world. As Bradley J. Fest argues, Negarestani "takes the challenge of thinking time in the era of climate change and, rather than jettisoning or evacuating narrative, finds that narrating through something else—that is, attempting to narrate a hyperobject, oil, rather than some human story situated amid a background of climate change—can produce a fiction appropriate for the realities of the Anthropocene" (Fest 2016, p. 573). Indeed, what *is* an appropriate fiction for the Anthropocene? *Cyclonopedia* makes no claim to accomplish this goal, but as Fest proposes, the centrality of oil as a narrative lubricant pushes the traditionally (white, male) human character aside—makes him "slip"—in a way that is quite refreshing. This is not to

[9] New materialism is an interdisciplinary field of inquiry interested in the intrinsic relationality of all matter and deeply critical of dualisms like human/nonhuman, nature/culture, and body/mind. It is associated with contemporary philosophers like Karen Barad, Jane Bennett, and Rosi Braidotti. As summed up in *The Posthuman Glossary* by Iris van der Tuin: "the new materialisms are mainly a research methodology for the non-dualistic study of the world within, beside and among us, the world that precedes, includes and exceeds us" (van der Tuin 2018, p. 277).

say that the book is lacking in (white, male) human influences. Negarestani provocatively meshes (and sometimes unapologetically plagiarizes) Lovecraft's and Deleuze and Guattari's ideas of cosmos, (non)chronology and a porous, vulnerable Earth. The thought experiments he throws the reader into are self-reflexive and uncomfortable, but the horror that crawls into the narration is cushioned by the dense philosophical jargon entangled with the political satire—even, at times, pure humor.

For example, in the first part of the book Negarestani paraphrases an online conversation between X and Z, who discuss the fictional *Defacing Ancient Persia* by Hamid Parsani, which theorizes oil as "an omnipresent entity" that functions as a "lube for the divergent lines of terrestrial narration" and "narrates the dynamics of planetary events" (Negarestani 2008, p. 19). Moreover, an ancient Middle Eastern cult called "the People of Naft (oil)," according to Parsani-via-Negarestani, has a list of the eleven monstrous qualities and sub-entities that stem from the "Ancient Enemy or Oil," number nine of which reads:

> The Pipeline-Crawler (Go-juice), a code name for an autonomous vehicle which smuggles Islamic war machines into Western Civilizations—but on the other side of the panorama, it is in fact the slow penetration of other narrative entities of petroleum into the rectal depths of all political orientations, whether formulated on religious platforms or not. Gas plays its role as an assistant culprit in making great distances accessible by applying pressure, pushing the flow to the furthest recesses of the globe. Petroleum is at the same time the desensitizer, the lubricant and the object of intrusion. (Negarestani 2008, p. 28)

The phallic image of pipelines penetrating into the "rectal depths of all political orientations" offers a hilarious caricature of fossil-fuel-loving politicians. The absurd humor of the passage might be explained by the "fact" that the conceptualization belongs to an oil cult's "promo-leftist conspiracy-mongering" (p. 19), but Negarestani has little patience for such rational explanations concerned with levels of fictionality. If anything, the passage leaves the reader more mystified than ever as to the line between fact and fabrication. The People of Naft are presented as quite transparently fictional, but their list's inclusion in the manuscript disputes one of Alvanson's observations from the foreword: "What is so bizarre about these writings—quite ironically, since I continue to read them—is that there is so much seething dread and outsized monstrosity prowling through the pages, yet there is no sex part, not even one penetration" (2008, p. xvi).

In fact, the word "penetration" is used several times in the text, most evocatively (and again, humorously), to describe the efforts of a mysterious indigenous people (the "Z crowd") in Ancient Iran to infiltrate the "Aryan" people who settled there (Negarestani 2008, p. 210). This last part of the book reads as a satirical re-inscription of Lovecraft's racist cosmic dread into Iranian (Aryanian) history, with the indigenous people of the Middle East representing a countercult(ure) to Lovecraft's recurring motif of mad cultists worshiping the Old Ones: "If, both for Lovecraft and the Aryans, purity must be safeguarded by an excessive paranoia, it is because only such paranoia and rigorous closure can attract the forces of the Outside and effectuate cosmic alienage in the form of radical openness—that is, being butchered and cracked open" (Negarestani 2008, p. 203). In other words, Negarestani suggests that Lovecraft's horror can be read in a way that circumvents and ridicules his "extreme racial paranoia" because this paranoia ironically attracts the Outside and "turns you into a fine meal, into a new meat ... a new food for a new earth" (200, ellipsis added). This seems like a bit of an overstatement of Lovecraft's affirmative potential, but Zach Blas, likewise, in a queer reading of *Cyclonopedia*, sees the book as "a love letter to the Outside," theorizing a "kind of love that goes beyond recognition, to a far more radical notion of otherness that the human cannot imagine" (Blas 2012b, p. 111; p. 112).

Radical openness, to Negarestani, is contrasted to "economic openness" or affordance, which is essentially a capitalist form of thought. To be radically open to the Outside is therefore a way of exiting human ("exhuming") religious, political, philosophical, and emotional frameworks so that even "[w]ide-open, open-minded, broadly-open and open-world, as the subjectively affirmed modes of openness, are rendered obsolete. Such spatio-logical manifestations of openness reinstall the logic of economical subjugations within the axioms of liberal commonsense" (Negarestani 2008, p. 201). This openness to exhumation is "always catastrophically unpleasant" for the subject (200), but ultimately, to Negarestani, it is productive because the goal is to "[d]isturb and irritate, dilate and contract the repressed cavities of the Earth: Tunnels and tubes, burrows and lairs, acrid bungholes and perforated spaces, its fanged vaginas, slits and the schizoid skin. Unclog and squeeze the earth; exhume its surfaces; make an earth whose conundrums cannot be solved by recourse to their origins or courses" (Negarestani 2008, p. 51). It is easy to see how critics like Blas might find productive dialogues between the growing field of queer studies, ecological thought, and Negarestani's radical departure from a heteronormative human perspective. This (queer/weird) departure also resonates with Seymour's *Bad Environmentalism*, which unmasks "the dominant preference for environmentalism to be straight, white, clean, and neat, despite the queer, diverse, messy grossness of the world, not to mention of environmental politics" (Seymour 2018, p. 38). Likewise, *Cyclonopedia* presents no systematic, neat approach to rethink the human in deep time. Indeed, in the hedged words of Fest, perhaps it can at best "concisely help us begin trying to represent the wildly differing scales of temporality necessary for placing the present within the now geologic history of the Anthropocene" (Fest 2016, p. 574). *Cyclonopedia*, in its leaps between horror and humor and its metafictional slips between theory and fiction, is perhaps most of all a celebration of "the queer, diverse, messy grossness of the world."

"Negarestani," writes Fest, "emphasizes how it is precisely literature—and experimental, difficult, and complex literature at that—that can open important, radical, twisted perspectives on life in the age of hyperobjects" (Fest 2016, p. 573). To Fest, *Cyclonopedia* unfolds "an imagination whose very despair exceeds the horror at the heart of contemporary crises" through its "rigorous nihilism": uninterested in apocalypse and species survival, the book claims that to "understand and confront life in the age of hyperobjects, thinking about deep time is an absolute necessity" (p. 575). And in its complex intermingling of fact and fiction, horror and humor, satire and seriousness, this is arguably the most important contribution *Cyclonopedia* makes to ecocritical thought—and to the weird. Negarestani's proposed strategy for opening the human imagination to a radical Outside is at once an intervention into sanctimonious Anthropocene philosophies obsessed with extinction and into the weird aesthetic of "serious" cosmic dread. Moreover, the hybrid, non-linear form of this prose-philosophy at once resists the realism of climate fiction (cf. Ghosh 2016, p. 32–33) and subverts the hegemony of the novel as a literary form.

Cyclonopedia relies heavily on Deleuze and Guattari as a stylistic and theoretical (almost theological) foundation, in particular their concept of holey space—or in Negarestani's bastardization, "( )hole complex": "Transpierce the mountains instead of scaling them, excavate the land instead of striating it, bore holes in space instead of keeping it smooth, turn the earth into Swiss cheese" (Deleuze and Guattari 1987, p. 413; Negarestani 2008, p. 41). ( )hole complex is the model for how oil narrates the Earth through its plot holes, based on compositions of solid and void and negotiations between them, so that "the solid narrates the anomalies generated by the void, or the infection of the void through the solid" (Negarestani 2008, p. 45). Perforating the surface of the Earth with (plot) holes, then, means "evoking periphery, itinerancy, and affect" and awakening "something radically irrelevant, having no correlation with its input, cause, or origin" (p. 45; 51). In these passages the meta is overtly present as a meta*phor* in the text itself, and the self-reflexivity is so potent that it makes little sense to separate meta from plot (holes) anymore. Rather, the activation of ( )hole complex relates to *both* the surface of the Earth and the surface of the text at the same time; arguably, the material brackets "( )" in front of "hole" also evoke the shape of a pipeline when viewed perpendicularly from one of the ends, looking through it like a tunnel.

The self-reflexive strategies and logical incongruities in *Cyclonopedia* create a fuzzy narrative space and slippery narrative time that ask a lot of readers. *Cyclonopedia*'s experimentation goes so far, however, that it might render purely narratological close readings moot. Logical incongruities, mysterious symbols, linguistic codes, dead URLs, and unfinished sentences poke deliberate holes in Negarestani's book, making the reading process more interactive and perhaps less immersive. As Ryan (via Douglas Hofstadter) suggests, these effects manifest themselves in the reading experience "as a strange loop: rising higher and higher through the levels of a hierarchical system, only to find oneself right where one began" (Ryan 2014, p. 807). However, Ryan goes on, "a narrative space cannot be wholly inconsistent, for fear of preventing any kind of mental representation—for fear, in other words, of losing its spatial quality. Logical contradiction is normally limited to some areas of the narrative world, piercing the fabric of space like the holes in a Swiss cheese" (Ryan 2014, p. 807; see also Hofstadter 1979, p. 686). Some might argue that *Cyclonopedia* is too inconsistent to live up to the minimal requirements of a fictional narrative because it has no real spatial setting. Rather, it travels nomadically from one plot hole to the next, regurgitating occult theories that connect the chapters thematically and affectively rather than spatially and temporally. However, the text still retains a sense of story because, in part, of the footnotes and preface.

Like holes in a Swiss cheese, the footnotes spread across the bottom of the pages (ostensibly following the order in which Alvanson finds them stuck to the manuscript) contain comments on the text by a person who might be Reza Negarestani, meant for someone who might be his publisher—or maybe Hamid Parsani. Or they might be the remarks of another reader, who, like Alvanson, has found the text and is "compelled to write on the left margin of a page of the manuscript" (p. xix). The "trippiest" scene in the footnotes arrives when the footnote narrator says their hotel room number is 302 and then describes a "blond woman in pink [*sic*] Death From Above T-shirt (didn't know they are this famous) and underwear in the hotel room across the garden, right in front of my window" (p. 37). Kristen Alvanson's room is also 302; she wears a "DFA 1979 T-shirt" and also complains that "someone in the window across the way" keeps looking at her (p. xvi). Like ( )hole complex and "petrological reason," the two characters are staged as "images of each other in an obsidian mirror" (Negarestani 2008, p. 43). The point of the footnotes, then, is to maximize the sense of mystery and activate readers as characters in the Cyclonopedic plot (or indeed: players in the Cyclonopedic game) at the same level as Alvanson and Negarestani. The ironic reference by the title to Wikipedia becomes palpable as readers (*this* reader, anyway) realize that they, too, have written notes in the margin related to fictionality (see Fig. 10.1).

Marshall suggests that *Cyclonopedia* has a lot to offer *as* a novel, precisely because it bends novelistic genre boundaries and asks readers to be

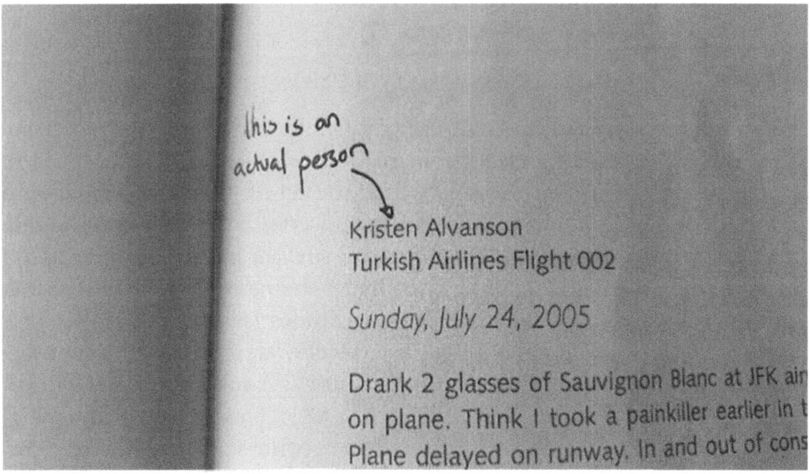

Fig. 10.1 "This is an actual person" (Alvanson/Ulstein, p. ix)

more than recipients, to interact and participate. But participate in what, exactly? To answer this, let us retrace the strange loop back to Alvanson's foreword, where we started, and where we find the second part of the definition of xenopoetics: "Fields of xenopoetics grow sporadically (until their final takeover of the work) around the regions overwhelming with a range of distortions from inauthenticity and corrupted authorship to structural holes (configurative bugs) and subterranean structures of hidden writing; xenopoetics does not necessarily insinuate adventurous modes of expression or prose" (Alvanson 2008, p. xix). In engaging with *Cyclonopedia*, readers participate in a xenopoetic mode of corrupted authorship and non-adventurous prose that takes them out of the traditional reader role and triggers their sense of co-authorship of those ideas, structures, and hierarchies that the book explores. It invites resistance to the hegemony of the novel (historically bound up in capitalism), both formally via the experimental strategies described above, and thematically, via the conceptualization of oil as narrative lube. Instead of being hostile to immersion, the text celebrates the power of weird fictionality to take readers, sliding on Tellurian Lube, into the plot holes of the perforated Earth. It invites readers to accept their complicity with anonymous materials like oil, to let go of the "hermeneutic rigor" of conventional reading and interpretation practices, and engage in the xenopoetics of Hidden Writing: "the ultimate tool of extraction, digging and participation, that is to say, reading as both scrutiny and realization" (Negarestani 2008, pp. 61–62).

10.4 Conclusion

The simultaneous scrutiny and realization mentioned in the quotation above is *Cyclonopedia*'s activation of readers as co-writers of the (story) world. Similarly, Erin James has proposed the act of writing as an analogical link between narrative and the Anthropocene, as "both narrative and the Anthropocene are products of ***humans writing worlds***" (James 2020, p. 187; emphasis added). According to James, humans "have rewritten the world to reflect their own activities and attitudes toward themselves and other species and matter. We literally see the effects of this rewriting in the inscription of human activity in geological strata. A similar sense of world-making is fundamental to narrative" (p. 188). This dual worldmaking power is central to *Cyclonopedia*, as readers become self-conscious of their participation as "characters" or "players" in the text and simultaneously awaken to their participation as characters or players in the Anthropocene world. But the worldmaking power is also visible in *Borne* on a thematic level, where it embodies a posthuman vision for a future in which nonhuman creatures will eventually "outstrip all of us in time, and the story of the [world] will soon be their story, not ours" (VanderMeer 2017, p. 317).

While *Borne* plays with narrative techniques to disrupt traditional weird tropes and challenge ecological thought, *Cyclonopedia* more elaborately plays with such techniques and engages the reader in a form of a politically loaded narrative "game" which is less about winning and more about realizing (and then resisting) the fact that the game has been rigged from the start by oil-infused processes like capitalism and petroculture. Both texts demonstrate an interest in moving away from normativity and neat or correct ways of imagining the Anthropocene—*Borne* on a thematic level and *Cyclonopedia* on a formal level as well. The two parts of this chapter have focused on two very different texts to suggest that the eco-weird affords forms and themes not typically associated with the weird as a literary genre or mode of writing. This affordance deserves more scholarly attention because, on the one hand, it accentuates the narratological components that have enabled the weird's trajectory from cosmic to a more explicitly environmental horror, which is useful for weird scholarship. On the other hand, it turns the focus toward minor, diverse, messy, and gross reactions to twenty-first-century ecological crises, which is important for the environmental discourse more broadly.

As mentioned above, Seymour suggests that it is worth paying attention to texts that "emphasize the nonnormative and refuse to route things back to the familiar" (Seymour 2018, p. 33). This is particularly prudent, she writes, "in a time when environmental crisis is widely normalized, whether as a coping mechanism or as a political strategy" (p. 33). As I have suggested in this chapter, contemporary eco-weird fiction is well-equipped for this kind of "bad environmentalist" work, especially because of the weird's evolution in the twenty-first century. Eco-weird texts experiment with unconventional techniques and less serious or sanctimonious themes and registers to highlight pressing questions related to human entanglement in and responsibility for nonhuman systems. They elaborately, joyfully, provocatively play with affects, nonhuman representation, metafictional and immersive strategies, and slip between horror and humor to find alternative, nonnormative angles to the cosmic horror of being human in deep time.

References

Alvanson, Kristen. 2008. Incognitum Hactenus. Foreword to *Cyclonopedia: Complicity with Anonymous Materials*. re.press, ix–xx.

Blas, Zach. 2012a. Hidden Writing. *American Book Review* 33 (3, March/April): 10–11. https://doi.org/10.1353/abr.2012.0062.

———. 2012b. Queerness, Openness. In *Leper Creativity: Cyclonopedia Symposium*, ed. Ed Keller et al., 101–114. Punctum Books.

Canavan, Gerry, and Andrew Hageman, ed. 2016. Introduction. "Global Weirding," special issue of *Paradoxa*, 28: 7–14.

Caracciolo, Marco, and Gry Ulstein. 2022. The Weird and the Meta in Jeff VanderMeer's Dead Astronauts. *Configurations* 30 (1), JHUP.

Carroll, Noël. 1999. Horror and Humor. *Aesthetics and Popular Culture*, special issue of *The Journal of Aesthetics and Art Criticism* 57 (2): 145–160, Wiley. https://www.jstor.org/stable/432305.

Clark, Timothy. 2020. Ecological Grief and Anthropocene Horror. *American Imago* 77 (1): 61–80, Project Muse. https://doi.org/10.1353/aim.2020.0003.

Deleuze, Gilles, and Felix Guattari. 1987. *A Thousand Plateaus: Capitalism and Schizophrenia*. Translated by Brian Massumi. University of Minnesota Press.

Economides, Louise, and Laura Shackelford, eds. 2021. Introduction. In *Weird Ecology: VanderMeer's Anthropocene Fiction. Surreal Entanglements: Essays on Jeff VanderMeer's Fiction*. Routledge, pp. 1–26.

Fest, Bradley J. 2016. Geologies of Finitude: The Deep Time of Twenty-First-Century Catastrophe in Don DeLillo's *Point Omega* and Reza Negarestani's *Cyclonopedia*. *Critique: Studies in Contemporary Fiction* 57 (5): 565–578. https://doi.org/10.1080/00111619.2016.1138446.

Fisher, Mark. 2016. *The Weird and the Eerie*. London: Repeater Books.

Ghosh, Amitav. 2016. *The Great Derangement: Climate Change and the Unthinkable*. University of Chicago Press.

Haraway, Donna J. 2016. *Staying with the Trouble: Making Kin in the Chthulucene*. Duke Press.

Harman, Graham. 2012. *Weird Realism: Lovecraft and Philosophy*. Zero Books.

Hegglund, Jon. 2020. Unnatural Narratology and Weird Realism in Jeff VanderMeer's Annihilation. In *Environment and Narrative: New Directions in Econarratology*, ed. E. James and E. Morel, 27–43. The Ohio State University Press.

Hofstadter, Douglas. 1979. *Gödel, Escher, Bach: An Eternal Golden Braid*. New York: Vintage Books.

James, Erin. 2020. Narrative in the Anthropocene. In *Environment and Narrative: New Directions in Econarratology*, ed. E. James and E. Morel, 183–202. The Ohio State University Press.

Lovecraft, H.P. 2013. Supernatural Horror in Literature. 1929. In *H. P. Lovecraft: The Classic Horror Stories*, ed. Roger Luckhurst, 444–447. Oxford University Press.

Luckhurst, Roger. 2017. The Weird: A Dis/Orientation. *Weird Fiction, Special Issue of Textual Practice* 31 (6): 1041–1061. https://doi.org/10.1080/0950236X.2017.1358690.

Marshall, Kate. 2012. *Cyclonopedia* as Novel (A Meditation on Complicity as Inauthenticity). In *Leper Creativity: Cyclonopedia Symposium*, ed. Ed Keller et al., 147–158. Punctum Books.

McGurl, Mark. 2012. The Posthuman Comedy. *Critical Inquiry* 38 (3, Spring): 533–553, University of Chicago Press, JSTOR. https://doi.org/10.1086/664550.

Miéville, China. 2015. Covehithe. In *Three Moments of an Explosion*, 337–351. Macmillan.

Morreall, John. 2020. Philosophy of Humor. In *The Stanford Encyclopedia of Philosophy*, ed. Edward N. Zalta, Fall 2020. www.plato.stanford.edu/archives/fall2020/entries/humor/.

Morton, Timothy. 2013. *Hyperobjects: Philosophy and Ecology after the End of the World*. University of Minnesota Press.

———. 2016. *Dark Ecology: For a Logic of Future Coexistence*. Columbia University Press.

Negarestani, Reza. 2008. *Cyclonopedia: Complicity with Anonymous Materials*. re.press.

Noys, Benjamin, and Timothy S. Murphy. 2016. Introduction: Old and New Weird. *Genre* 49 (2): 117–134, Duke University Press. https://doi.org/10.1215/00166928-3512285.

Ryan, Marie-Laure. 2001. *Narrative as Virtual Reality: Immersion and Interactivity in Literature*. The Johns Hopkins University Press.

———. 2014. Space. In *Handbook of Narratology*, vol. 1, ed. Peter Hühn et al., De Gruyter, 796–811.

Said, Edward W. 1993. *Culture and Imperialism*. Vintage Books.

Seymour, Nicole. 2018. *Bad Environmentalism: Irony and Irreverence in the Ecological Age*. University of Minnesota Press.

Strugatsky, Arkady, and Boris Strugatsky. 2012. *Roadside Picnic*. 1972. Translated by Olena Bormashenko, Chicago Review Press.

Swanson, Heather et al., ed. 2017. Introduction: Bodies Tumbled into Bodies. In *Arts of Living on a Damaged Planet: Ghosts of the Anthropocene*, M1–M12, University of Minnesota Press.

Tan, Shaun. 2020a. *A Bear and Her Lawyer*. www.shauntan.net/bear-print. Accessed 05 January 2024.

———. 2020b. Bears with Lawyers. *Fabula*, special issue of *Dark Mountain*, no. 18, edited by Nick Hunt et al., 125–129.

Traub, Courtney. 2018. From the Grotesque to Nuclear-Age Precedents: The Modes and Meanings of Cli-Fi Humor. *Studies in the Novel* 50 (1): 86–107. Project Muse. https://doi.org/10.1353/sdn.2018.0005.

Ulstein, Gry. 2019. What's Weird about Giant Flying Bears? *Collateral*, no. 15, collateral-journal.com/index.php?cluster=15.

———. 2021a. Heights They Should Never Have Scaled: Our (Weird) Planet. *SUB-STANCE* 50 (3): 14–33. https://doi.org/10.1353/sub.2021.0020.

———. 2021b. *Weird Fiction in a Warming World: A Reading Strategy for the Anthropocene*. Ghent University. Faculty of Arts and Philosophy. http://hdl.handle.net/1854/LU-8729679.

———. 2021c. "Just a Surface": Anamorphic Perspective and Nonhuman Narration in Jeff VanderMeer's *The Strange Bird*. In *Nonhuman Agencies in the Twenty-First-Century Anglophone Novel*, edited by Liebermann, Yvonne, Judith Rahn, and Bettina Burger, 277–296. Palgrave Macmillan. https://doi.org/10.1007/978-3-030-79442-2_14.

Van der Tuin, Iris. 2018. New Materialism. In *Posthuman Glossary*, edited by Rosi Braidotti and Maria Hlavajova, 277–279. Bloomsbury Academic.

VanderMeer, Jeff. 2010. Weirdly Funny. www.jeffvandermeer.com/2010/03/04/weirdly-funny/. Accessed 16 July 2021.

———. 2017. *Borne*. MCD: Farrar, Straus and Giroux.

———. 2019. *Dead Astronauts*. Farrar, Straus and Giroux.

Vermeulen, Pieter. 2020. *Literature and the Anthropocene*. Routledge.

Waters, Colin N., et al. 2016. The Anthropocene Is Functionally and Stratigraphically Distinct from the Holocene. *Science* 6269 (351). https://doi.org/10.1126/science.aad2622.

Index[1]

[1] Note: Page numbers followed by 'n' refer to notes.

B. H. Onishi, N. M. Bell (eds.), *The Call of the Eco-Weird in Fiction, Films, and Games*,
https://doi.org/10.1007/978-3-031-77126-2

The manufacturer's authorised representative in the EU is Springer
Nature Customer Service Centre GmbH, Europaplatz 3, 69115 Heidelberg,
Germany. If you have any concerns regarding our products, please
contact ProductSafety@springernature.com

Printed and bound by CPI Group (UK) Ltd, Croydon, CR0 4YY
01/10/2026
02225953-0008